DISTRIBUTION

Scale "AA" arms concerned—Manual of Military Pubns.

HANDBOOK
OF
ENEMY AMMUNITION

CONTENTS TABLE

German Ammunition Markings and Nomenclature

LIST OF PLATES

HANDBOOK OF ENEMY AMMUNITION

PAMPHLET No. 15

GERMAN AMMUNITION MARKINGS AND NOMENCLATURE

Introduction

The German ammunition authorities started the war with a systematic ammunition marking code, but owing to rapid development of ammunition, the introduction of substitutes and the carry over of old stocks, the system has, in some cases, been extended and become confused.

Also, there are to be found in the German service specimens of ammunition of French, Czech, Polish, Italian, etc., manufacture. These are still being used with their original markings which bear no relation to the German system and may have certain German markings added to them.

In addition, air and naval service ammunition is extensively used in the Army, the marking of which differs somewhat from that of the Army.

This pamphlet gives the marking and nomenclature that has been met in the course of examination of captured specimens and documents, and has been divided into two sections.

Section I deals with Small Arms Ammunition.

Section II deals with Gun Ammunition.

For obvious reasons Section I deals with individual types of Small Arms Ammunition whereas Section II deals with Gun Ammunition in a more general form.

There is also a note on Mortar Ammunition, and at Appendix G, a list of abbreviations used in the nomenclature of grenades.

SECTION I

SMALL ARMS AMMUNITION

The small arms ammunition dealt with in this pamphlet is sub-divided under the following headings :—

SUB-SECTION A.

(*a*) 7.92 mm. Rifle and M.G. ammunition (Ground use).
(*b*) 7.92 mm. Rifle and M.G. ammunition (Miscellaneous).
(*c*) 7.92 mm. M.G. ammunition (Aircraft guns).
(*d*) 7.92 mm. Rifle Grenade Propelling Cartridges.
(*e*) Pistol and Machine Carbine ammunition.
(*f*) Anti-tank Rifle ammunition.

SUB-SECTION B.

(*a*) 20 mm. (Oerlikon Flak) ammunition (2 cm. A.A. gun—Flak 28).
(*b*) 20 mm. (Solothurn) ammunition (2 cm. A.A. and A. tk. and 2 cm. Tank Guns—Flak 30 and 38 and Kw.K).
(*c*) 13 mm. (Solothurn) ammunition (Aircraft M.G. 131 El).
(*d*) 15 mm. (Mauser) ammunition (Aircraft M.G. 151 and 151 El).
(*e*) 20 mm. (Oerlikon) ammunition (Aircraft M.G.—FF and FFM).
(*f*) 20 mm. (Mauser) ammunition (Aircraft M.G. 151/20 and 151/20 El).
(*g*) 30 mm. (Solothurn) ammunition (3 cm. Aircraft guns—Mk. 101 and 103).
(*h*) 30 mm. (Mk. 108) ammunition (3 cm. Aircraft gun—Mk. 108).

ABBREVIATIONS

A list of abbreviations and their meanings, will be found at Appendix A. The list is in alphabetical order and is sufficient to assist in the identification of small arms ammunition.

SUB-SECTION A.

1. Packing

The standard method of packing small calibre ammunition is in the following containers :—

(*a*) small cardboard cartons (Faltschachtel) which vary in size according to the calibre of the round.
packed in :—
(*b*) cardboard carriers (Packhülse 88).
packed in :—
(*c*) ammunition boxes (Patronenkasten 88) or zinc lined ammunition boxes (luftdichte Patronenkasten 88).

Details :—

Ammunition Type	Carton Contents	Total	Carrier Contents	Total	Ammunition Box Contents	Total
7.92 mm. Rifle	3 chargers with 5 rounds in each charger	15	20 cartons	300	5 carriers	1,500
7.92 mm. M.G.	Loose rounds	15	20 cartons	300	5 carriers	1,500
Bulletted blank	Do. (or) loose rounds	15 50	Do.	300	Do. 29 cartons	1,500 1,450
Propelling Cartridges	issued one	with	each Rifle Gre	nade		
9 mm. Pistol	Loose rounds	16	52 cartons	832	5 carriers	4,160
7.92 mm. (Pist. Patr. 43)	Loose rounds or in loading chargers, 5 rounds per charger	14 or 15 or 20	30 cartons 30 ,, 22 ,, (In future the into the amm	420 450 440 cartons unition	5 carriers 5 ,, 5 ,, will be packed box—not in the	2,100 2,250 2,200 direct carriers)
7.92/13 mm. A.tk.	Loose rounds	5			50 cartons	250

2. Package Labels

Specimens of the standard package label are illustrated on Plate I.

The top line or main heading on the label gives the general identification of the ammunition. (The large label on the ammunition box also gives the total number of rounds contained).

Other information given on the label :—

Type of propellant.

Details of manufacture of the components and the complete round.

To assist in quick recognition the following system of markings and colours is also used :—

"Für Scharfschützen geeignet." marked on box.	"For snipers" Ball (m.s. core) selected in manufacture.
i.L. stencilled on in large red letters.	Rounds packed in chargers for the rifle.
Ex. stencilled on in red.	Drill rounds.

White label with black prints.	Normal ball ammunition.
"S.m.K." of the top line printed in red.	A.P. ammunition.
Top line in white letters on a black background.	Incendiary ammunition.

A blue vertical band.	Steel cartridge cases.
A green diagonal band.	Practice (A.A.) ammunition.

A yellow label.	Tracer.
A brown or pink label.	Bulletted blank.
A blue label.	9 mm. pistol ammunition.
A green label ("P.m.K." of the top line printed in red).	A.P. incendiary.
A diagonally half red/half white label.	A.P. (Tungsten carbide core).

3. Cartridge case markings

There are four items stamped on cartridge bases :—
Two items are found on all rounds :—

1. Year of manufacture, e.g. 42.
2. Manufacturing batch number, e.g. 56.

The other two items may be any two of the following :—

3. Code name of manufacturer—a combination of three small letters, e.g. aux. or single capital letter, e.g. P.
4. Type of cartridge case, e.g. S* (= brass).
   St, St+ } (= steel).
   roman numerals, etc., e.g. IXwI (= brass coated steel case).
5. Calibre e.g. (for the MP43)—7,9.

4. Colour markings on the round (Plates II to IV)

The system of identification by colour marking on the ammunition, varies from one calibre to another. The tables in this pamphlet should be consulted for definite identification. In the range of 7.92 mm. Rifle and M.G. ammunition the following general rules apply :—

Green or blue cap or cap annulus	=	Ball.
Red cap or cap annulus	=	A.P.
Black cap or cap annulus	=	Incendiary.
Black bullet tip	=	Tracer.
A green ring round the bullet (or green tip to bullet)	=	Higher velocity for aircraft guns.

5. Materials

Cartridge cases are now all made of steel, protected by a grey-green lacquer.

During the change over from brass to steel, some cartridge cases were made of steel, protected by plating with copper or brass.

Steel cases have also been thinly coated with wax in an attempt to overcome the hard extraction problem with the M.G. 42 at higher temperatures.

The percussion cap which was of brass is now being made of steel, zinc plated.

The bullet envelope is made of steel, normally clad G.M. (Gilding Metal). Zinc coating in place of gilding metal is known however to have been introduced recently for two types.

	Type	Bullet Type	Remarks	Identification
	(*a*) **7.92 mm. Rifle and M.G. (Ground Use)**	**Plate II**		
1	Ball (s.S.)	Lead core Steel (clad G.M.) envelope	Being replaced by the mild steel core round	Green cap annulus.
2	Ball (m.s. core) (S.m.E.)	Mild steel core. Lead sleeve. Steel (clad G.M. or zinc) envelope	Now used as the standard ball ammunition	Blue cap or cap annulus.
3	Ball (m.s. core) (long) (S.m.E. lang)	Mild steel core. Small lead-ring sleeve. Steel (clad G.M. or zinc) envelope	(Overall length unchanged). Bullet lengthened to compensate for loss of weight due to small lead content, and set deeper in the cartridge case	Colourless varnish on the zinc cap.
4	A.P. (S.m.K.)	Flat base hard steel core. Lead sleeve. Steel (clad G.M.) envelope	Also used in aircraft guns	Red cap or cap annulus.
*5	AP/T (S.m.K. L'spur)	ditto with tracer cup and base washer	ditto Usually yellow trace Some have green/red trace Some have green trace	Red cap or cap annulus black tip (10 mm.) to bullet.
6	Low Velocity Ball (Nahpatrone)	As for ball	For use with the rifle fitted with a silencer	Bright green lacquered case.
	(*b*) **7.92 mm. Rifle and M.G. (Miscellaneous)**	**Plate II**		
7	A.P. (Tungsten) (S.m.K. H.)	Tungsten carbide core. Lead sleeve Steel (clad G.M.) envelope	Obsolescent	Red cap or cap annulus. Black bullet.
8	Explosive Incendiary (B.)	White phosphorus in nose. Lead plug containing fuze. Steel (clad G.M.) envelope	Serves as an observing round when used by ground troops Also used in aircraft guns	Black cap or cap annulus. Black bullet with plain tip (early types—plain bullet with a chromium plated tip.)

	Type	Bullet Type	Remarks	Identification
9	Practice Ball (le.S.)	Aluminium core Steel (clad G.M.) envelope	Practice A.A. light bullet	Green stripe on cartridge base.
10	Practice Tracer (le.S. L'spur)	Aluminium core Tracer cup Steel (clad G.M.) envelope	ditto	Green stripe on cartridge base. Black tip (10 mm.) to bullet.
11	Bulletted Blank (platz)	Hollow wooden bullet	For firing M.Gs. on exercises	Black cap or cap annulus. Purple wooden bullet.
	(*c*) **7.92 mm. (for aircraft machine guns only) Plate III**			
12	High Velocity A.P. (S.m.K.v.)	As 4 above	Was used with slightly modified M.G. 17 Not recently encountered	Red cap or cap annulus. Green ring 10 mm. from bullet tip (early type had green bullet tip).
*13	AP/T 100/600 Trace (S.m.K. L'spur 100/600)	As 5 above	Dark ignition for 100 metres Duration of trace 600 metres	Red cap or cap annulus. Black tip (10 mm.) to bullet.
14	H.V. AP/T 100/600 Trace (S.m.K. L'spur 100/600 v)	As 5 above	*See* 12	Red cap or cap annulus. Black tip (10 mm.) to bullet. Green ring 10 mm. from bullet tip.
15	AP/T night tracer (S.m.K. Gl'spur)	As 5 above		Red cap or cap annulus. Black tip (5 mm.) to bullet.
16	H.V. AP/T night tracer (S.m.K. Gl'spur v)	As 5 above	*See* 12	Red cap or cap annulus. Black tip (5 mm.) to bullet. Green ring 10 mm. from bullet tip.
17	AP/I (P.m.K.)	Shaped steel core Lead nose and base plug. White phosphorus surrounds core stem Steel (clad G.M.) envelope		Black cap or cap annulus (Occasionally red annulus, or red band).

* NOTE.—Nos. 5 and 13. The only method of distinguishing between these two types is by the package label.

	Type	Bullet Type	Remarks	Identification
18	H.V. AP/I (P.m.K. v)	ditto	*See* 12	Black cap or cap annulus. Green ring 10mm. from bullet tip (early type had green tip.)
19	H.V. Explosive Incendiary (B-v)	As 8 above	*See* 12	Black cap or cap annulus. Black bullet, plain tip Green ring 10 mm. from bullet tip.

(*d*) **7.92 mm. Rifle Grenade Propelling Cartridges. Plate IV.**

	Type	Type of Cartridge Case	Remarks	Identification
1*	For the propaganda grenade (G. Kart für G. Propgr)	Crimped mouth		Red cap annulus.
2	For the H.E. Grenade (G. Kart für G. Sprgr)	Crimped mouth	Early pattern	Yellow cap annulus.
3*	For the H.E. Grenade (G. Kart für G. Sprgr)	Long neck closed by a plug held in by coning the mouth	Replaces (2)	Yellow cap annulus.
4	For the H.E. Grenade (Treibpatr. für G. Sprgr)	Closed with a blue wooden bullet	Being introduced as stocks of (3) are used	Blue wooden bullet.
5*	For the H.E. Grenade (long range)	Closed with a yellow wooden bullet	To boost the grenade range (from 250) to 500 yards approx.	Yellow wooden bullet.
6*	For the small A.P. Grenade (G. Kart. für G. Pzgr.)	Crimped mouth		Black cap annulus.
7*	For the large A.P. Grenade (G. Treibpatr. für gr.G. Pzgr)	Closed with a black wooden bullet		Black wooden bullet.
8	For the small A.P. Grenade SS.46	Crimped mouth		Plain brass cap.
9	For the large A.P. Grenade SS.61	Long neck closed by a plug held in by coning the mouth		Plain zinc cap.

* Illustrated on Plate IV

	Type	Bullet Type	Remarks	Identification
	(*e*) **Pistol and Machine Carbine Ammunition. Plate IV**			
1	7.92 mm. Ball (Pist. Patr 43)	Mild steel core, lead sleeve Steel (clad G.M.) envelope		Zinc cap plain lacquered, or blue cap annulus.
2	9 mm. Ball (Pist. Patr 08)	Lead core Steel (clad G.M.) envelope	Brass case Obsolescent	Black cap annulus.
3	9 mm. Semi AP (Pist. Patr 08 m.E.)	Mild steel core Lead sleeve and base Steel (clad G.M.) envelope	Becoming the standard ball	Black cap annulus (was black bullet now plain bullet).
4	9 mm. Ball (sintered iron) (Pist. Patr. 08 m.S.E.)	Solid sintered iron		Grey coloured bullet. Black ring at junction of bullet and case.
5	9 mm. Low velocity ball (Pist Nahpatrone 08.S.)	Lead core	Bullet 3 mm. longer to increase weight. Overall length of round unchanged	Bright green lacquered case. Plain bullet.
	(*f*) **Anti-Tank Rifle Ammunition (7.92 mm./13 mm.). Plate IV**			
1	Practice Ball	Lead core Steel (clad G.M.) envelope	Intended for range practice (Not illustrated)	Plain bullet. Green cap annulus. Stamped △ on cartridge base.
2	AP/T (lachrymatory) (S.m.K. H.Rs. L'Spur) Labelled : Patronen 318	Tungsten carbide core Lead sleeve G.M. envelope Tracer cup and lachrymatory pellet		Black tip to bullet. Red cap annulus. Brass case.
3	Propelling Cartridge for the A. tk. Rifle Grenade (Treibpatrone 318)	Mouth closed with plain wooden bullet		Plain wooden bullet.

Sub-section B. (13 mm. to 30mm. calibre).

As there is a large variety of shell types this pamphlet has been limited to details of :—

20 mm. Oerlikon (Flak)	A.A. and A.tk. ammunition.
20 mm. Solothurn	

and reference only to :—

13 mm. Solothurn	Aircraft gun ammunition.
15 mm. Mauser	
20 mm. Oerlikon	
20 mm. Mauser	
30 mm. Solothurn	
30 mm. Mk. 108	

The general description of packing, labels and colour identification given below, applies to the *two land service types of ammunition.* There are variations of the system in the air service ammunition and there are one or two instances where colours on the shell conflict with this general system. Accurate identification depends upon correct translation of the package label.

1. Packing

All types are issued unbelted. Each round is contained in a cardboard cylinder closed at one end to protect the nose or nose fuze of the shell.

These cylinders are packed in wooden, zinc lined ammunition boxes. This zinc lining is sealed air-tight in three ways—soldered, or by means of a lid, seated in rubber or luting.

2. Package labels

The label has a main heading which gives the designation of the ammunition either in full or in abbreviation.

Other information on the label gives the manufacturing codes and some detail of the components.

For quick recognition, the following colour system is used on the label :—

All yellow label	Indicates	H.E.
Yellow label with red triangle in each corner	,,	H.E./I.
White label over-stencilled with a red "O"	,,	A.P. (inert or no filling).
White label over-stencilled with a red "Ph" ...	,,	A.P./I.
White label over-stencilled with a red "Z"	,,	A.P./H.E.(S.D.).
Diagonally half white/half red label	,,	A.P. (tungsten carbide core).
Vertical blue band	,,	Steel cartridge cases.
Diagonal green band ...	,,	Practice ammunition.

Additional letters in the main designation on the label :—

(Oerl) indicates Oerlikon ammunition for the Flak 28.

'Flak' indicates the types of Solothurn ammunition which are for A.A. only.

3. Markings of rounds (Plate V)

(*a*) *Shell colours :*

A yellow shell with black stencilling	=	H.E.
,, ,, ,, red ,,	=	H.E./I.
black ,, ,, white ,,	=	A.P.
grey ,, ,, black ,,	=	Practice.
bright unpainted alloy or black painted, pointed shell	=	A.P. (tungsten carbide core).

(*b*) *Coloured bands on the shell*

A 6 mm. coloured band above the driving band indicates the colour of the tracer.

(A yellow tracer is not indicated on a yellow shell).

A thin red ring above the driving band (superimposed on the tracer band if necessary) indicates tropical loading. In this case a red tracer is indicated by a red band below the fuze or a red tip on A.P. shells.

A black ring above the driving band is sometimes used to indicate the absence of self-destruction on H.E. shell.

(*c*) *Stencilling on the shell*

The following stencillings are common to all types :—

Manufacturer's code	e.g. avu.
Delivery code number ...	e.g. 2a.
Year of manufacture	e.g. 41.
Weight of shell in grammes...	e.g. 148g.

Additional information (e.g. W, Z, Br., etc.) appear on one or two rounds and are mentioned specifically in the tables. (*See* list of abbreviations for their meaning).

NOTE :—

20 *mm. types :*

There are four types of 20 mm. ammunition. External views are shown on Plate VI, and as will be seen they can be identified by the dimensions and shape of the cartridge case. All cases originally brass are now made of steel (lacquered).

4. Notes on Land Service Ammunition (the identifying feature of the label designation is in italics).

(*a*) 20 *mm. Oerlikon* (*Flak*)

The case is rimless ($1\frac{1}{2}$ in. longer than the air service type) and has a shoulder. The ammunition is percussion fired.

Example of the designation on the package label :—

2 cm. Brand-Sprenggranat Patronen 40 L'Spur (rot) (*Oerl*).

The following types have been encountered :—

20 mm. Oerlikon—Flak

Type	German Nomenclature	Colour of Projectile	Colour Bands on Projectile	Stencilling
HE/T (S.D.)	2 cm. Sprenggranat-Patrone 40 L'spur (rot) (Oerl)	Yellow aluminium fuze	Red band above driving band	Normal stencilling and " W " in black.
HE/I/T (S.D.)	2 cm. Brand-Sprenggranat-Patrone 40 L'spur (rot) (Oerl)	Yellow aluminium fuze	Wide red band also covering the driving band	Normal stencilling and " W " and " Br " in red.
HE/I (S.D.)	2 cm. Brand-Sprenggranat-Patrone ohne L'spur (Oerl)	Yellow aluminium fuze	—	Normal stencilling and " Br. o L'spur " in red.

(*b*) 20 *mm. Solothurn*

The case is belted (the only one in the 20 mm. group) and is the longest of the 20 mm. group. The ammunition is percussion fired.

Example of the designation on the package label :—2 *cm.* Br-Sprgr-Patronen L'Spur (rot).

(NOTE :—No identifying letters or numbers after the heading).

The following types have been encountered :—

20 mm. Solothurn

Type	German Nomenclature	Colour of Projectile	Colour Bands on Projectile (as encountered)	Stencilling (other than normal)
HE Practice	2 cm. Sprenggranatpatrone (Üb)	Grey		Black arrow on body
HE/T Practice	2 cm. Sprenggranatpatrone L'spur (Üb)	Grey Dummy aluminium fuze	Yellow band above driving band. (D.B.)	Black arrow on body
AP Practice	2 cm. Panzergranatpatrone (Üb)	Grey		Black arrow on body
AP/T Practice	2 cm. Panzergranatpatrone L'spur (Üb)	Grey	Colour Band above D.B. indicates the colour of the tracer	Black arrow on body

Type	German Nomenclature	Colour of Projectile	Colour Bands on Projectile (as encountered)	Stencilling (other than normal)
HE/T (S.D.)	2 cm. Sprenggranatpatrone L'spur (gelb) or (rot) or (weiss)	Yellow alluminium fuze	Above D.B. :— No band = yellow tracer Red band = red tracer White band = white tracer	
HE/T (S.D.)	2 cm. Sprenggranatpatrone L'spur W	Yellow aluminium fuze	Ditto	" W " in black
HE/I/T (S.D.)	2 cm. Brand-Sprenggranatpatrone L'spur (Flak)	Yellow aluminium fuze	Some have black band above D.B.	" W " and " Br " in red
HE/I/T (S.D.)	2 cm. Brand-Sprenggranatpatrone vk. L'spur (Flak)	Yellow aluminium fuze	Colour band above D.B. indicates the colour of the tracer	" Br vk L'spur " in red
HE/I/T (S.D.)	2 cm. Brand-Sprenggranatpatrone vk. L'spur W (Flak)	Yellow aluminium fuze	Colour band above D.B. indicates the colour of the tracer	" W " and " vk L'spur " in red
HE/I (S.D.)	2 cm. Brand-Sprenggranatpatrone o L'spur (Flak)	Yellow aluminium fuze		" Br o L'spur " in red
AP/T	2 cm. Panzergranatpatrone L'spur	Black	Colour band above D.B. indicates the colour of the tracer	" O " in white = empty or inert filled
AP/I/T	2 cm. Panzergranatpatrone L'spur	Black	Yellow band above D.B. or red tip and red band above D.B.	" Ph " in white
AP/T (Tungsten carbide)	2 cm. Panzergranatpatron 40 L'spur	Light alloy or Black	Red Band above D.B.	
AP/HE/T (S.D.)	2 cm. Panzergranatpatron L'spur (gelb or rot) mit Zerlegung	Black	Yellow or red band above D.B.	" Z " or " ZERL " in white

5. Notes on Air Service Ammunition. (The identifying feature of the label designation is in italics.)

(*a*) 13 *mm. Solothurn*

The cartridge case, originally of brass, is now made of steel coated copper or brass and is the belted type. All types are electrically fired. None of the shells is self-destroying.

Example of the designation on the package label :—

13 mm. Brsprgr. Patr. L'spur El. o. Zerl.

(*b*) 15 *mm. Mauser*

The cartridge, originally brass, is now made of steel (lacquered) and is the rimless type and has a shoulder. All types are percussion fired. (A German document has mentioned electrically fired 15 mm. Mauser, but none have been encountered).

Example of the designation on the package label :—

15 mm. Pzgr. Patr. L.spur o. Zerl.

(*c*) 20 *mm. Oerlikon* (Air service)

The case is rimless (the shorter of the two Oerlikon types) and has no shoulder. The ammunition is percussion fired.

Example of the designation on the package label :—

2 cm. Brsprgr. Patr. L'spur *FF*. o. Zerl.

or

2 cm. M.Gesch. Patr. *FFM* m. Zerl.

(*d*) 20 *mm. Mauser*

The case is rimless, about the same length as Oerlikon (Air Service) but larger diameter and has a shoulder. Both percussion fired and electrically fired types are issued.

Example of the designation on the package label :—

2 cm. M-Gesch. Patr. *151* m. Zerl.

2 cm. Brsprgr.-Patr. L'spur *151 El.* m. Zerl.

(*e*) 30 *mm. Solothurn*

The case is belted, is very long, with a short neck and shoulders. The rounds are percussion fired and those intended for the Mk. 103 are electrically fired.

Example of the designation on the package label :—

3 cm. Sprgr. Patr. L'spur o. Zerl.

or

3 cm. Sprgr. Patr. *El* o. Zerl.

(*f*) 30 *mm. M.K.*108

The case is short in comparison with the length of the projectile. It is rimless, and is made of steel coated inside and out with G.M. or brass. The ammunition is electrically fired.

Example of the designation on the package label :—

3 cm. Sprenggranat-Patronen *MK 108 El.*

SECTION II

GUN AMMUNITION

General

German gun ammunition is named according to four different systems.

(*a*) Nomenclature by calibre and type of shell (e.g. 10 cm Gr. 19).

(*b*) Nomenclature by type of gun and nature of shell (e.g. F.H. Gr.).

(*c*) Nomenclature by construction of the shell (e.g. 15 cm Hbgr. 16 umg.).

(*d*) Nomenclature by description of the shell (e.g. 8.8 cm Sprgr. L/4.5).

In addition, the abbreviation Patr. (meaning Q.F. fixed) sometimes followed by the name of the gun which the cartridge fits is added to the nomenclature of fixed ammunition.

Of the above systems (*a*) and (*b*) are the current German army systems, (*c*) being obsolescent and (*d*) being used mainly for Naval or Air service shell.

German gun ammunition is almost all Q.F., there being only one gun of German manufacture which fires B.L. ammunition. Ammunition for Anti-Aircraft, Tank and Anti-Tank guns is in the "Q.F. fixed" class, the remainder with few exceptions (e.g. 7.5 cm FK. 38) is in the "Q.F. separate" class.

In some cases the calibre given in the nomenclature is nominal and not the actual calibre of the gun; for instance, the 10.5 cm medium gun is known as the "s.10 cm K. 18."

A list of abbreviations commonly found in the nomenclature of German ammunition together with their German and English meaning is shown in Appendix B.

The list does not include complex abbreviations used by the Germans which are built up from the simple basic abbreviations. These will be found listed under their separate parts.

For typical package labels *see* Plate XVII.

MARKINGS ON CARTRIDGE CASES AND CHARGES

(Plates VII to XII)

Cartridge Cases (General)

The information given on the cartridge case is in the form of stencilling on the side and base of the case and stampings on the base. The stencilling may be in black or white, except when it relates to propellant charges for use in hot climates in which case it is in red.

GERMAN FIXED Q.F. CARTRIDGES.
TYPICAL MARKINGS.
Case lot.
Year of manufacture of case.
Case maker.
Shell.
Sprgr.
44
P.150
1938
6347
8·8 cm. Flak18
Calibre, type & model of equipment.
Nature & model No. of projectile—A-P. with T.-C. core. (Stencil).
Cartridge case.
Pzgr.40
Steel case.
Va
41
6360 St.
7·5 cm. Pak
40
Model No. of case.
Anti-tank gun – Model 40.
Calibre & type of equipment.
Date of manufacture of primer.
Primer maker.
22
Primer lot.
Charge weight.
aue
40
Primer No.
Nature, shape & size of propellant.
3·7cm.Pak.
164 g.
8354St.
7·5 cm. KwK
Smoke shell (Stencil).
Place & date of manufacture of propellant with workmark.
Tg.J.1940/19
H.E. Shell (Stencil).
Sprgr.
14
Weight class of projectile. (Stencil).
aux
42
Ce.I.24I.A.
Place & date of filling.
P.T. +25°C.
MwP
6339St.
7·5 cm. KwK40
Standard charge temperature (Hot climate charge).
Tank gun—Model 40.

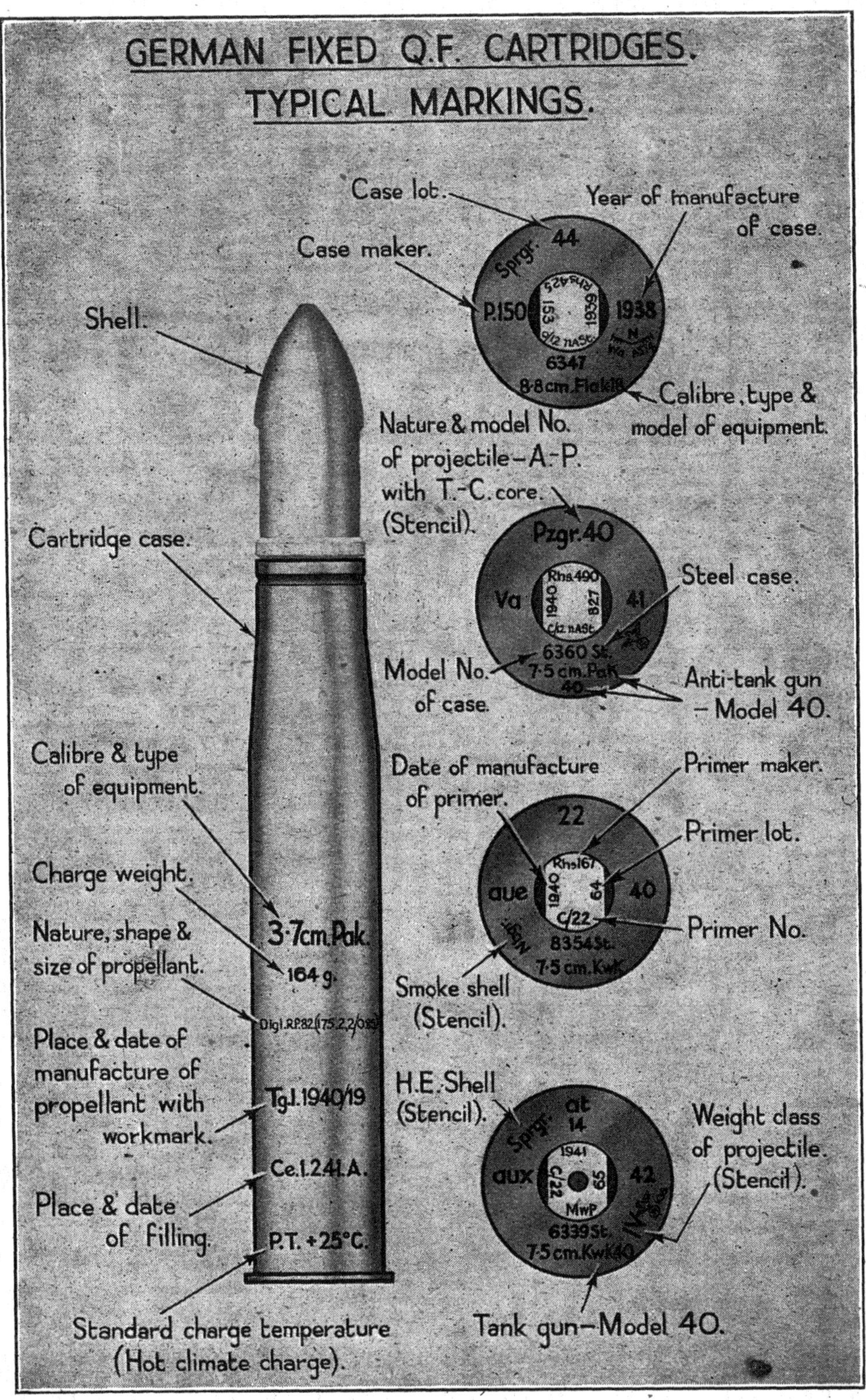

Fig. 1

Stencilling on the side of the case. (Fig. 1)

The following details are arranged in the sequence in which the markings are usually found.

(*a*) Designation of Gun, i.e. calibre, type and model of the equipment for which the round is suitable.
e.g. 7.5 cm Kw.K.40 meaning 7.5 cm. tank gun model 40.

Where a round is suitable for more than one equipment the designations are stencilled in sequence.

When the equipment is of foreign origin a small letter in brackets, e.g. (*t*) for Czech, (*p*) for Polish, etc., is added to the designation, normally after the model number.

(*b*) Actual weight of propellant in grams, e.g. " 164 g," or if the weight is over 1,000 grams it is usually indicated in Kilograms e.g. " 1,225 Kg."

(*c*) Nature, shape and size of propellant. (*See* Appendix C).

(i) Nature. This is given in the form of a code which is an abbreviation of the German description of the composition ; for example " Ngl or NG " is the abbreviation of the word " Nitroglyzerinpulver," the composition being a double base propellant of nitroglycerine and nitrocellulose. These letters, in the case of double base propellant, are followed by figures or letters which appear to relate to the composition calorific value and percentage of potassium sulphate incorporated.

(ii) Shape. This is given in abbreviated form followed by the letter P, which is the initial letter of Pulver (Powder).
e.g. Bl.P. = Blättchenpulver, meaning Rectangular flakes.
A list of the abbreviations will be found in Appendix C.

(iii) The size of the propellant is given by a statement of the dimensions in millimetres following the letters used to indicate the shape. The dimension figures are enclosed in a bracket and are arranged as follows with commas serving as decimal points, and full stops or crosses (x) separating the individual figures.

Flake	(length . breadth . thickness) e.g. (3 · 3 · 0,8).
Circular washers ...	(thickness, external diameter/internal diameter), e.g. (1,9 · 15/4).
Tubular	(length with minus tolerance, external diameter/internal diameter), e.g. (175-2, 2/0,85).
Strip	(length . breadth . thickness), e.g. (125 · 5 · 0,5).
Multi-perforated Discs	(Diameter . thickness), e.g. (50 · 0,2).

Chopped Cord ... (Length . Diameter), e.g. (1,5 · 1,5).

Granular ... ... (Min. Diameter—Max. Diameter.), e.g. (0,3 - 1,5).

The following are typical examples of the complete markings used to indicate the nature, shape and size :—

" Digl.R.P. - 8,2 - (175 - 2,2/0,85)."
" Gu.Bl.P. - AO - (4 · 4 · 0,6)."
" Ngl.Bl.P. - 12,5 - (40 x 40 x 0,2)."
" Nz.R.P. - (135 · 5,5/2)."
" Digl. Str.P. - 9,2 - (125 x 5 x 0,5)."

(*d*) Three letter code of propellant factory, year of manufacture of the propellant, and delivery number.
e.g. " dbg 1942/3."

(*e*) Place, date of filling and work mark.
e.g. " On 17.6.42.V."

(*f*) Details of standard charge temperature or tropical loading.

Red stencilling is used to indicate propellant charges of a reduced weight for hot climates and may be found near the base of the case, just above the flange, or higher up the side of the case, above the other stencilling. The marking used, " P.T. + 25° C.," indicates that the normal or standard charge temperature on which the weight of the charge is based is 25° C. (i.e. 77° F.). The German standard charge temperature for normal European temperatures is 10° C. (i.e. 50° F.).

In some instances cases are stencilled " Abgebr Ldg " in red. This marking is found near the base (corresponding to the position of the " P.T. + 25 deg. C." marking) and indicates that the weight of the propellant charge has been reduced for use in hot climates. The standard temperature on which the reduction is based is indicated by the stencilling " Schusstafeln P.T. + 50° C.", also in red. This standard was superseded by that of 25° C.

Where the propellant charge is suitable for use in hot climates but the charge weight is adjusted for the standard charge temperature of 10° C. the cases are stencilled " Auch für Tropen."

Stencilling on the Base of the Case (Fig. 1)

The positions of the following markings are as viewed with the case turned so that the stamped letters and numbers on the bases are upright.

Distinctive markings in script lettering which indicate the nature, and in some instances the model number, of the projectile are stencilled in white or black paint to the left above the primer hole. The markings used to indicate the type of shell are the same as those given above under " Projectile Nomenclature."

In some instances the Roman numerals indicating the weight classification of the projectile are stencilled in white to the right below the primer hole.

Stampings on the Base of the Case (Fig. 1)

These form a record of certain particulars regarding the cartridge case.

Fig. 1 shows the normal position of the base stampings and gives their significance. The design number (case model No.) below the primer hole, is followed by the letters " St. " when the case is of steel.

The abbreviations following the calibre of the gun stamped below the design number of the case are the same as those given in the details of the stencilling on the case.

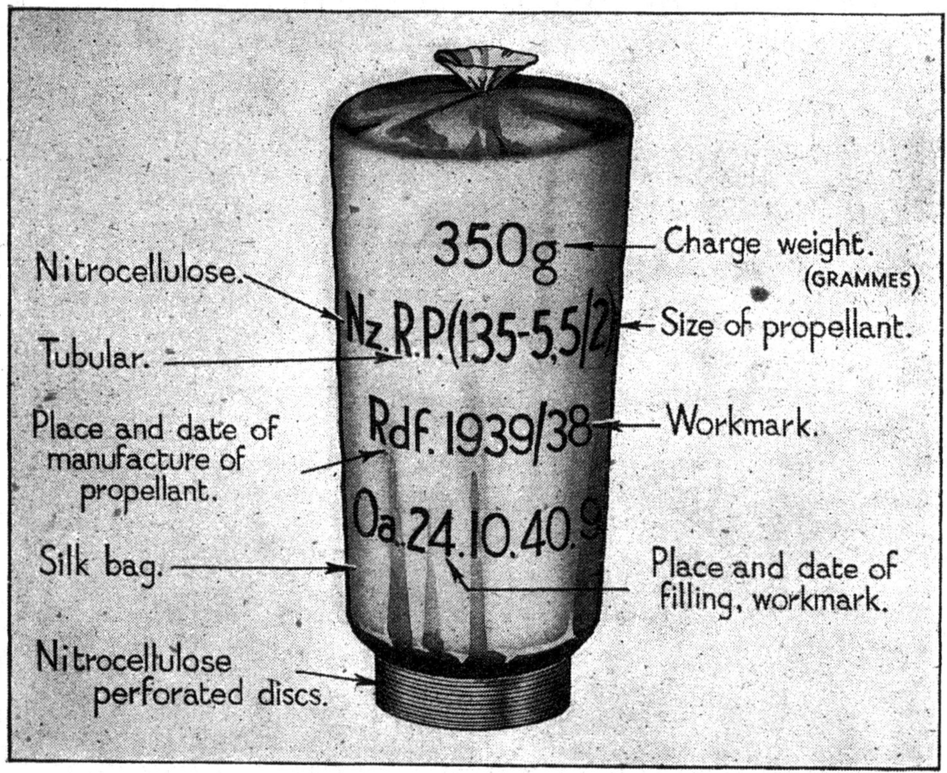

Fig 2

Markings on Charge Bags

The markings on the bags are the same as those stencilled on the side of the case except that the calibre, type and model number of the equipment are not always included.

A typical example is given in Fig. 2.

BASE OF CASE.
P
23
1937
6350
sF.H.18.
CARDBOARD CLOSING CUP.
s.F.H.18.
Ktz. 1940/32
Le 29.4.41 F
Für Tropen!!
+ 25°C.
CHARGE SECTION.
6D
s.F.H.18.
475g
Digl B.I.P. 10,5(4x4x1,2)
Ktz. 1940/32
Le 28.4.41 F
Bleidraht im Beutel
Black lettering.
Red lettering.
Silk bag.
Lead wire decoppering agent included.

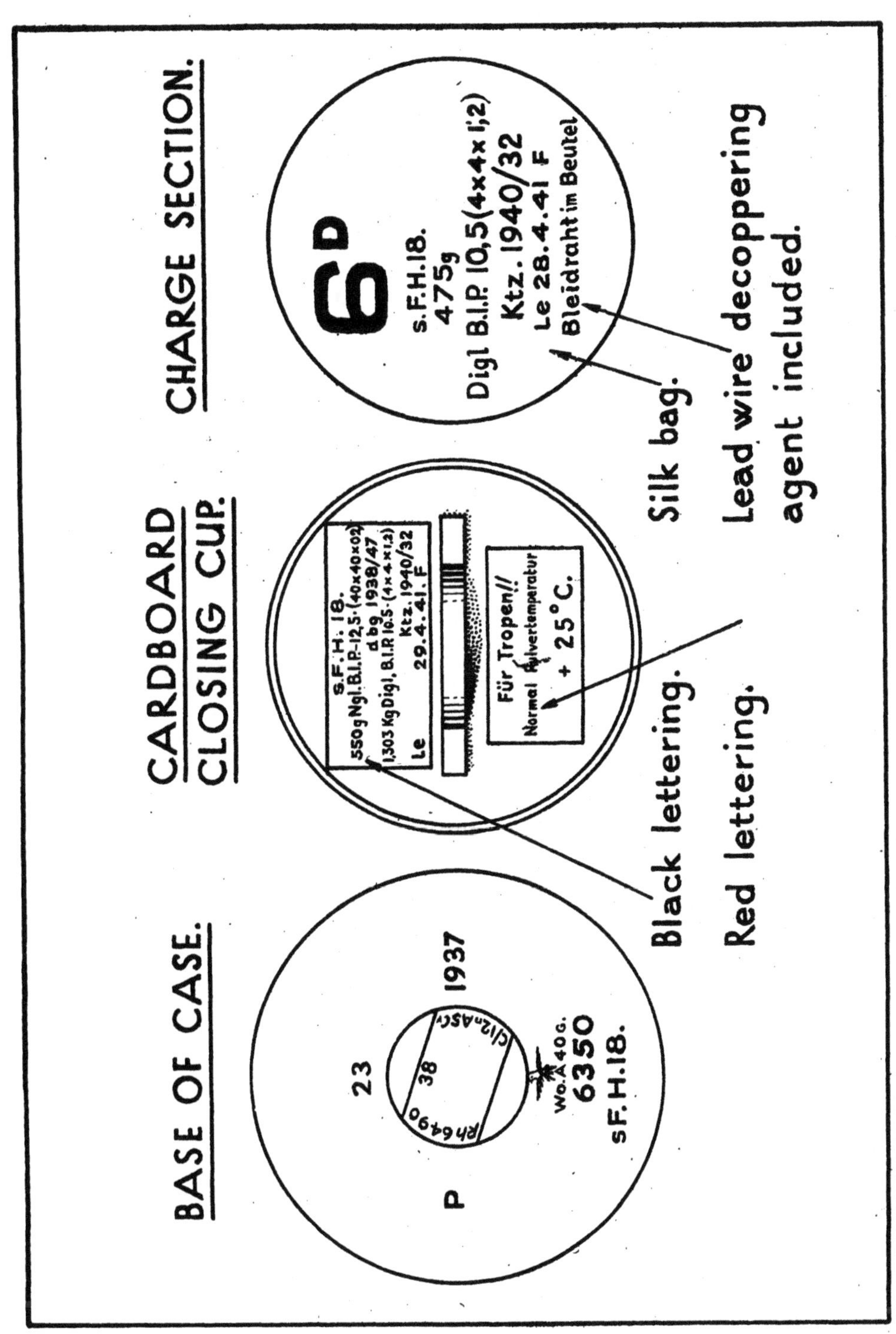

Fig. 3

SEPARATE LOADING AMMUNITION

Stencilling and Stamping (Figs. 3 and 4)

The stencillings on the case follow the same procedure as that for "fixed ammunition."

The stampings on the base of the case are also the same as those on the base of the fixed round except that the calibre of the gun is sometimes omitted and that additional markings may be used.

The following are some additional abbreviations in the stamped designation.

Stamping	Indication
Geb. H. (Model No.) ...	Mountain Howitzer.
Geb. K. (Model No.) ...	Mountain gun.
le. F.H.	Light field howitzer (British equivalent, gun howitzer).
s.F.H.	Heavy field howitzer (British equivalent, medium howitzer).
le. I.G.	Light infantry howitzer.
s.I.G.	Heavy infantry howitzer.
s. (Calibre) K	Heavy gun (British equivalent, medium gun).
L.G.	Light gun, recoilless.
K (Model No.) (E) ...	Railway gun.

The abbreviations used for howitzers are also used for gun-howitzers. The model number which follows the abbreviation differentiates between these types.

The design number followed by a second number separated by an oblique stroke, followed by a letter, denotes a built up case. While it is not known what the second number stands for, the letter stands for the type of built up case.

Closing Cups and Covers (Fig. 3)

Where cardboard or leatherboard cups are used to close the mouth of the case, labels on the cup, printed in black, give details corresponding to those stencilled on the side of fixed Q.F. cases, i.e.

(*a*) Designation of the equipment.
(*b*) Charge weight in grams or kilograms.
(*c*) Nature, shape and size of the propellant.
(*d*) Place and date of manufacture.
(*e*) Place and date of filling.

The label indicating propellant charges for hot climates with a charge weight based on a normal charge temperature of 25 deg. C. is printed in red.

Cases with steel covers for packing and transport, which are removed before loading, have neither labels nor stencilling relating to the propellant charge, except the stencilling "P.T. + 25° C". in red on the base when applicable. Details of the propellant are available, however, from the stencilling on the charge bags.

Silk bag.
Charge Section Nº
1
Standard charge temperature (hot climate charge).
BASE OF CASE.
3
Case lot.
PT+25°C
P
1936
6349
S 10 cm K18
Maker of case.
Case model Nº.
Date of manufacture of case.
Equipment calibre & model number.
Equipment.
Charge wt.
S.10 cm K.18
2375 Kg
Digl. R.P. GO.5(450-31/1,S)
Nature shape & size of propellant.
tgl 1940/1
Be.18.5.41 K.
PT.+25°C
Place & date of manufacture of propellant.
Date of filling.
IGNITER ON BASE.
S.10 cm K.18
4 OgNzManNP(15
rtl 1940/8
Be 5.4.41K
Standard charge temperature.

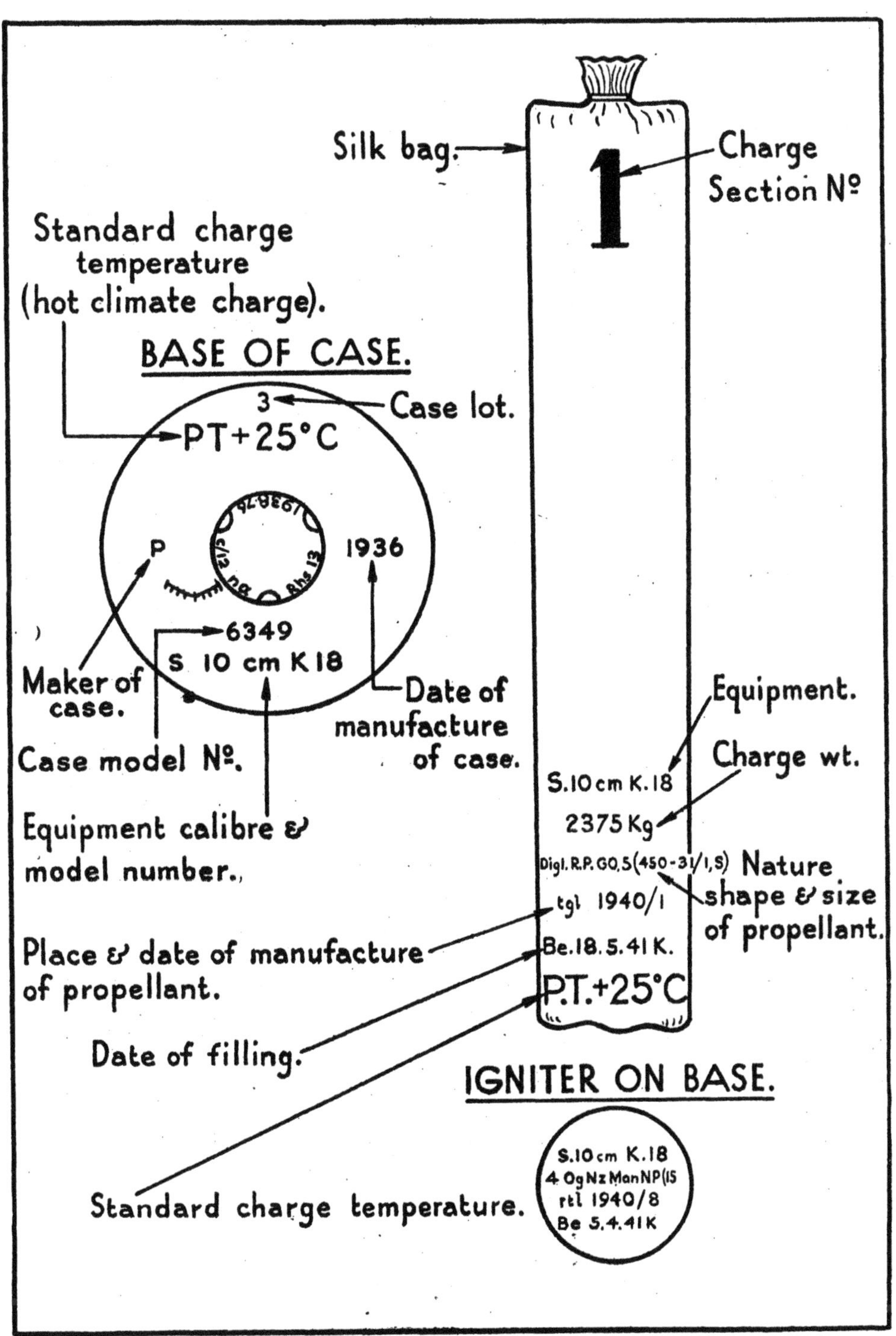

Fig. 4

Charge Bags (Plates VII and VIII)

With separate loading ammunition the German nomenclature relating to charges is based on the following terms.

Hülsenkart	meaning	separate loading cartridge (normal charge).
Hauptkart	,,	main charge.
Vorkart	,,	incremental charge used with Hauptkart.
Sonderkart	,,	substitute charge.
Grundladung	,,	basic or bottom section.
Teilkart	,,	part charge.
Kart-Vorl	,,	flash reducing charge.

All except the " Kart Vorl " may be fitted with igniters.

Howitzer charges are usually confined to Teilkart and Sonderkart but an unusual combination of all of them is made use of in forming gun charges.

The build up of howitzer charges follows very much the same methods used in the British service, e.g.

Charge 1 = 1 Ladung = Teilkart 1
Charge 2 = 2 Ladung = Teilkart 1 + 2.
Charge 5 = 5 Ladung = Teilkart 1 + 2 + 3 + 4 + 5
etc.

In some cases the higher charges comprise one or more substitute (Sonderkart) charges used without the " Teilkart " portions. The sequence of numbering remains unaltered, e.g.

6 Ladung = Teilkart 1 + 2 + 3 + 4 + 5 + 6
7 Ladung = Sonderkart 7.

NOTE.—The Sonderkart is a complete charge in itself and is packed as a separate item. The charge numeral is enclosed in a red ring which denotes that it is not to be used with the other part charges without red rings.

Guns are generally provided with three charges, the build up of which varies considerably. In a few instances they are straightforward as for howitzers, i.e. Teilkart 1 + 2 + 3 but the majority are invariably made up into combinations of the above and are known as kleine (small), mittelere (medium) and grosse (large) charges before loading according to tactical requirements.

The method of packing also varies. In some instances the smaller charges are issued in the cartridge case and the larger charges packed in containers, in others the medium and large charges are issued in their own cartridge cases and so on.

The following are examples of known charge combinations taken from captured German documents.

Ex. 1	kl. Ladung	=	Sonderkart 1.
(15 cm. K. 18)	mittl. Ladung	=	Hauptkart + Vorkart 2.
	gr. Ladung	=	Hauptkart + Vorkart 2 + Vorkart 3.

Ex. 2	kl. Ladung	= Grundladung + Sonderkart 1.
(15 cm. K. 16)	mittl. Ladung	= Grundladung + Teilkart 2.
	gr. Ladung	= Grundladung + Teilkart 2 + 3.
Ex. 3	kl. Ladung	= Grundladung + Teilkart 1.
s. 10 cm. K. 18	mittl. Ladung	= Grundladung + Teilkart 1 + 2.
	gr. Ladung	= Sonderkart 3.

The markings on the charge bags consist of the following information (*see* Figs. 3 and 4).

(*a*) Code letter for type of propellant (*see* Plate XVIII).
(*b*) Number or name of part charge.
(*c*) The designation of the gun for which the charge is made up.
(*d*) Actual weight of propellant in grams or kilograms.
(*e*) Type and size of propellant.
(*f*) Code letter for place of production, year of production and lot number.
(*g*) Code letter for place of filling, date of filling and work mark of the filling factory.
(*h*) Details of standard charge temperature or tropical loading.

When a decoppering agent is included in the charge, the bag is marked "Bleidraht im Beutel." (Fig. 3).

Flash Reducing Charges (*See* Plate VIII)

These are enclosed in flat circular silk bags which may be stencilled in black "Kart. Vorl" followed by the abbreviation indicating the equipment in which it is to be used, and the weight of the charge in grams, or marked with the weight and chemical formula, e.g. 30 g. K_2SO_4 or not marked at all, in which case the fact that it exists is given elsewhere on the case or package.

Additional Abbreviations and Markings found in connection with Charges

Marking	Where found	Meaning
o B D m B D	On the cartridge case or cartridge cup	Charges without } decoppering foil. ,, with
Red Ring	Round body of charge bag	Special part charge to be used only with other red marked part charges.
Umgesetzt followed by place initials, date and trade mark	Cartridge closing cup	Charge bags transferred to a new cartridge case with date, trade mark, etc., of factory making the change.
R	On the charge bag	Charge to be used with rocket assisted shell only.
U	Cartridge cup	Cartridge has been examined.
F	On the charge bag	Charge to be used with long range shell only.
M or Man	On the cartridge case	Blank cartridge.

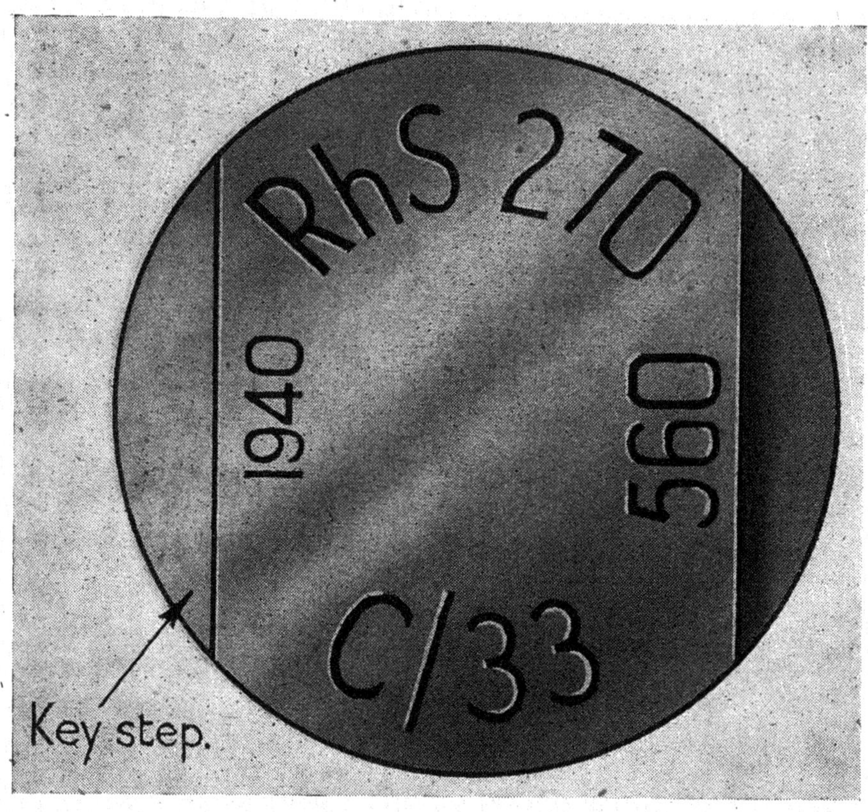

Fig 5

PRIMERS

Marking of Primers for Q.F. Cartridges

Primers are normally designated in a numbered series with the letter " C " and an oblique stroke immediately in front of the number. The letters "n.A." and " St." (in dicating new pattern and steel respectively) are included in the designation of those in common use.

The following are known primers. Typical stampings are shown in Fig. 5.

(*a*) Percussion : enclosed cap and anvil plug.

C/12 nA	Standard primer for 5 cm and above. (Pamphlet No. 4, p. 10).
C/13 nA	Smaller calibres 2.8 cm, 3.7 cm, etc. (Pamphlet No. 7, p. 44).
C/33	Small brass primer (Pamphlet No. 7, p. 45).
C/43	Same size as C/13 nA but has thinner diaphragm under cap. Used with 7.5 cm L.G.40.

(*b*) Electric : used with tank guns and 8.8 cm Flak 41, 8.8 cm Pak 43, etc.

GERMAN SHELL STAMPINGS.
Fuze.
Acceptance stamp (1st test).
Delivery number; firm; year of manufacture.
Acceptance stamp water pressure test.
Mark of assembly firm.
Acceptance stamp (2nd test).
Shell model number.
Acceptance stamp (hardness).
Acceptance stamp of release.
Acceptance stamp (release).
Projectile model number.
Delivery number; firm; year of manufacture.
Acceptance stamp.
Acceptance stamp for fitted base.

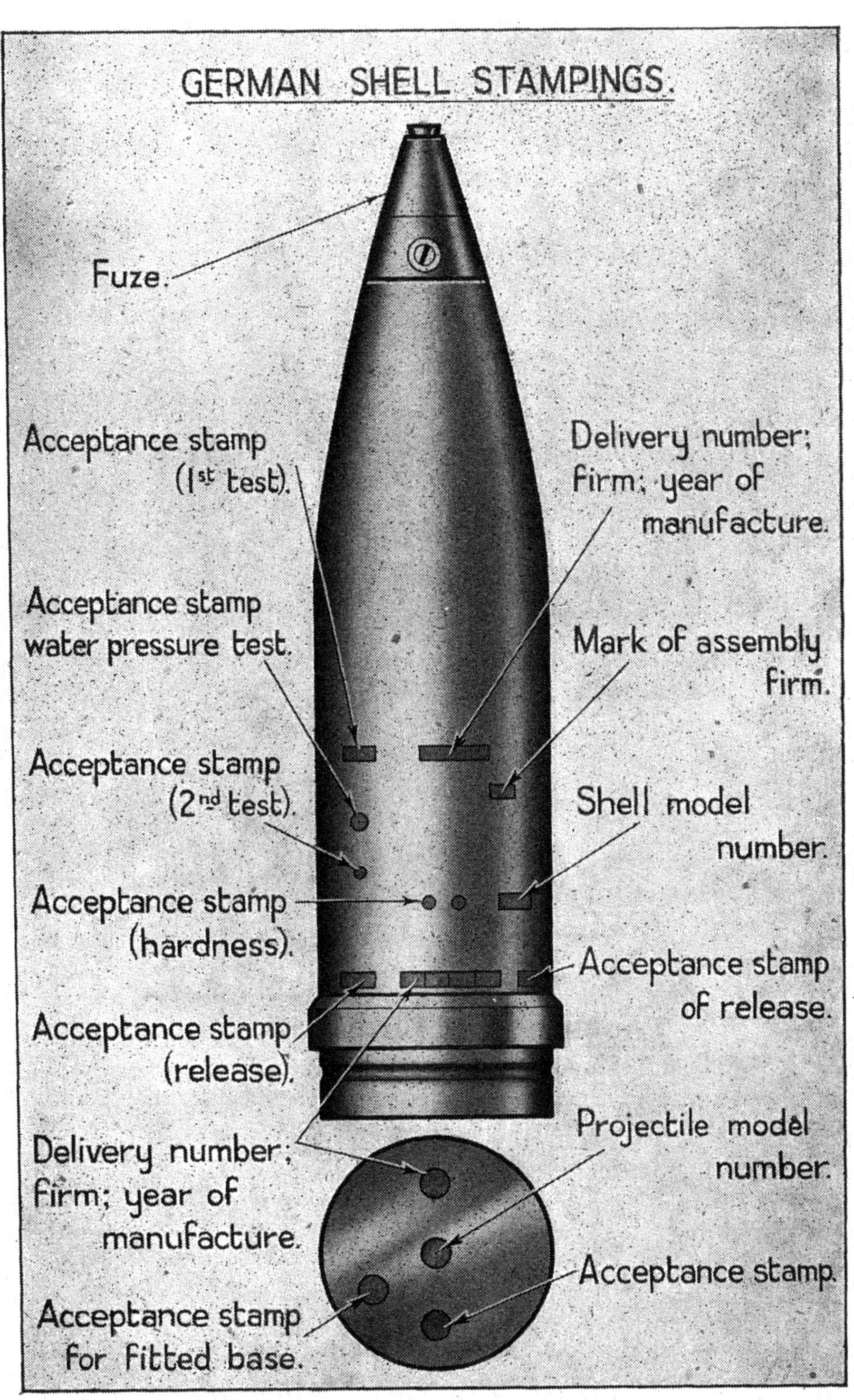

Fig. 6

C/22 Same gauge as, and interchangeable with, C/12, e.g., 5 cm. Pak 38 and Kw.K 39 same amn. after changing primer. (Pamphlet No. 4, pp. 30, 31).

C/23 Small type. Used with spigot mortar bombs and amn. for 3.7 cm Kw.K.

PROJECTILES

(Plates X to XIV)

Nomenclature

All types of projectiles, including mortar bombs, grenades and rockets usually have the abbreviation "Gr" or "gr" meaning Granate, or for certain non-explosive projectiles (e.g. Star) "G" or "Gs" meaning Geschoss, included in their basic designation. In addition to the calibre, the abbreviation may have a prefix to indicate the type of projectile or the equipment in which it is used, e.g. "Sprgr" for H.E. shell, "Wgr." for Mortar, "Igr" for Infantry gun, etc. A suffix consisting of numerals (Model No.) or letters or both, is included in the designation of projectiles which are different materially, ballistically or in action, from another of the same calibre, and is particularly important for identification purposes, when they closely resemble each other. The following are examples :—

8 *cm. Mortar Bombs*

Wgr. 34... ... A normal type of bomb with the head integral with the body.

Wgr. 38... ... A jumping bomb. The head which is lightly attached to the body accommodates an ejection charge to throw the bomb into the air after impact.

Wgr. 38 umg. ... The above modified to convert it into a normal bomb to burst on impact.

In the case of some A.P. projectiles the number provides a means of distinguishing shot and shell, and whether they possess a penetrative cap when fitted with ballistic caps. For example, the following shell have the same external appearance.

7.5 *cm Pak or Kw.K.* 40

Pzgr 39 = A.P.C.B.C. *Shell.*

Pzgr 40 = A.P.B.C. *Shot.* Has a T.C. core and is without a penetrative cap.

The letters, which normally follow the number, indicate the character or purpose of an individual type, e.g.

15 cm Gr. 19 Be = 15 cm. Model 19, Anti-concrete shell.

15 cm Gr. 19 Nb = 15 cm. Model 19, Smoke shell.

In this connection too, shell of foreign origin and non-German calibre have the calibre in millimetres marked on them.

GERMAN SHELL MARKINGS
TYPICAL STENCILLING.
Fuze.
13
Nature of filling.
Place and date of fuzing with workmark.
Lr. 9.6.41. G.
IV
Weight classification.
Cast steel shell.
Stg
Place and date of filling with workmark.
Copper and iron driving band.
KPS
Lr. 4.6.41. L.

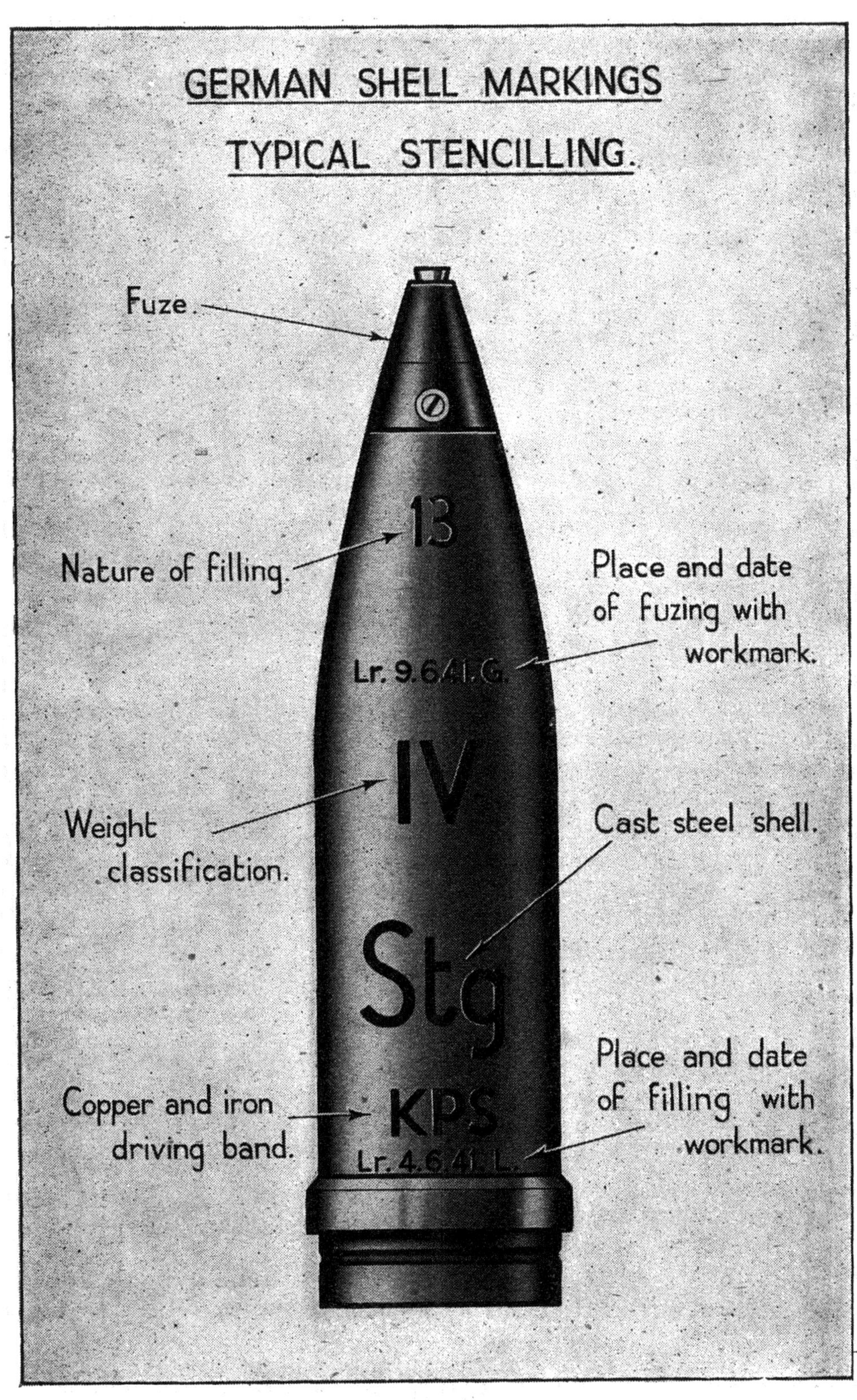

Fig. 7

The above designations are not as a general rule stencilled on the shell of "fixed" Q.F. ammunition but may be found on the base of the cartridge case and the base of separate loading shell. They will invariably be found on packages and package labels.

Further abbreviations are given in the Appendices.

Markings

Shell marking is, in general, very uninformative, only those markings which are necessary for the gunner or inspector are given. For this reason a very close examination is necessary to identify it.

Shell markings may be divided into two main classes :—

1. Stampings.
2. Painting and Stencilling.

Stampings

Stampings on shell are mostly manufacturing or inspection marks, and for recognition purposes may be completely ignored.

Fig. 6 gives the normal positions and the significance of the stampings on shell.

Stencilling

Stencilling is the most important consideration in the examination of shell with a view to identification.

All stencilling on shell, other than A.P. types, is normally in black. Stencilling on A.P. types is normally in red.

Fig. 7 and the plates at the end of this pamphlet are typical examples of the more general and important stencilling which will be found on most shell. Appendix D explains others in use.

For the purpose of identification, the shell exterior may be divided into the following areas. (*See* Fig. 8).

A. Immediately below the fuze, or the tip in the case of A.P. shell.
B. The ogive.
C. The cylindrical portion of the body.
D. Immediately above the driving band.
E. The side below the driving band.
F. The base of the shell.

In area A is stencilled in black the initials of the place where the fuze is inserted in the shell together with the date and identification mark of the firm. Details affecting the fuzing are also stencilled here. Shell with ballistic caps, fitted with nose fuzes, are invariably stencilled on the cap with the abbreviated designation of the fuze.

In area B is stencilled in large black arabic numerals the code number denoting the nature of the bursting charge. Appendix E gives the known code numbers. Below the filling number, and in larger Roman numerals, is stencilled the classification of the shell

AREAS OF SYSTEMATIC SHELL MARKINGS
A
B
C
D
E
She 2944 SKb
13 A
II
oR
NG 13.2.43 F FES
BASE
MARKINGS
FES
15cm.Gr. 36 FES
Dopp Z. S/60s
II

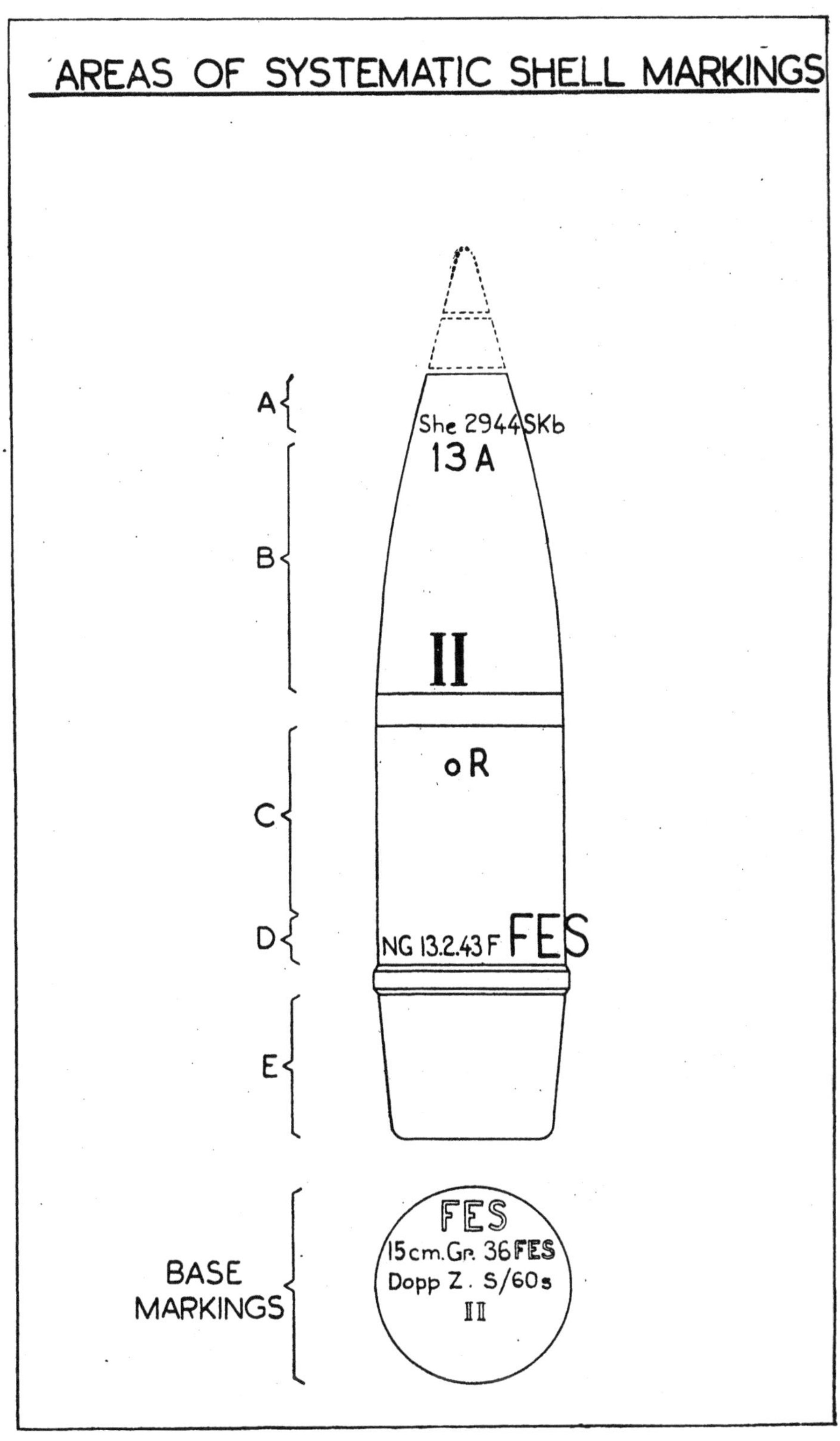

Fig 8.

for weight. Shell are classified in five weights, I to V. Those within " dead weight limits " are stencilled III, the lightest I and the heaviest V. The stencilling is in 1.2 inch numerals for calibres up to 21 cm and in 2 inch numerals for shell of larger calibre. One or both of the above numerals may appear in area C.

In *area C* is stencilled all the details concerning the type of shell and its action. On smoke shell, the date and trade mark of the filling factory is stencilled beside the filling plug, if such exists, otherwise this information appears in area D. Various stampings will also be found in this area which refer to manufacturing details affecting the shell and driving band.

In *area D* is stencilled the initials of the place, date and trade mark of the shell filling factory. In addition any details of the driving band which the user should know, though this may overflow into area " C."

Stampings also appear here. These are :—Delivery No., shell manufacturers initials or monogram, year of manufacture and a work mark.

In *area E* (separate loading type only) the shell calibre is sometimes stencilled in white, but usually there is no stencilling in this area.

The base is sometimes stencilled with a brief resume of the important details on the side of the shell, e.g. type of driving band, calibre and nature of shell, type of fuze, weight class.

PAINTING

Basic colour.

The bodies of projectiles are painted as follows. Depth of colour may vary slightly :—

Black	Armour Piercing shot and shell.
Olive Green, Olive drab, or Field Grey	H.E. Shell (except A.A. and Naval types), smoke shell, anti-concrete shell, hollow charge shell, and sea ranging A.P. shell.
Yellow	A.A. and Naval H.E. shell.
Blue	Naval A.P. shell.
Aluminium ...	Some shell (other than C.W. types) with coloured bands.
White	Shell in experimental stage.
Pale Green ...	Naval Star shell.
Dull Red or Deep Olive Green ...	Mortar bombs.
White above shoulder Red below	Propaganda shell.
Orange	Naval H.E./I shell.

Tips of Projectiles

The painting of tips is not a common practice as with British projectiles and has no general significance, i.e. it has a different meaning with individual natures.

A white tip is painted on 7.62 cm. shell to enable the ammunition to be readily identified as such. The cartridge case of this round is interchangeable with, and bears markings relating to, the 7.5 cm Pak 40 ammunition.

A white tip on 8.8 cm A.P.C.B.C. projectiles denotes a low capacity filling.

A black tip on Naval H.E. shell, which are painted yellow, indicates filled shell.

Bands around the body

White band (area C) ... Position of the centre of gravity on shell above 28 cm.

Yellow band (area C) ... To differentiate between two projectiles of the same calibre which are similar in appearance but differ materially in ballistic performance.

Yellow band (area D) ... On a "7.5 cm K Gr rot KPS" Shell immediately above the driving band has a dual meaning. It denotes the shell has a modified driving band (usually indicated by a red band in this position) and its ballistic performance differs from another shell of similar appearance (Yellow band area C).

The range tables for above shell are marked with a yellow stripe.

Red band (area D) ... Used to distinguish a projectile with a modified driving band where two otherwise identical type shell exist.

Coloured band (area D) ... On 2 cm, and some naval shell, the colour of the band indicates the colour of the trace.

White band (area E) ... Below the cannelure of fixed Q.F. ammunition to indicate the position of the cannelure has been changed.

Crosses

White cross ... Percussion fuze only to be used.

Black cross ... Time fuze only to be used.

Yellow cross ... Time and percussion fuze only to be used.

Red cross ... Gaine fitted to shell.

Miscellaneous Markings

Perpendicular stripes ... Modifications incorporated in the shell.

Black arrow pointing towards the nose or the base of the shell. On naval yellow coloured shell with ballistic cap indicates nose or base fuze according to direction of arrow.

FUZES

(Plate XV)

Fuzes are either issued unpainted, painted or rustproofed according to the material of which they are made. For example, fuzes made of aluminium, brass and plastic are not normally painted, whereas steel fuzes are painted, usually the same colour as the shell in which they are used, or rustproofed.

Loose fuzes are not normally issued as a separate component to be fitted in the field.

The nomenclature of nose fuzes is usually stamped on the body just above the edge which butts against the wall of the shell or adapter and includes the date of manufacture and trade mark of the manufacturer. Base fuzes are usually stamped in a similar manner on the exposed surface of the base of the fuze.

Percussion Fuzes

Percussion fuzes are generally designated by a serial number following letters which denote the type of fuze, e.g. " A.Z.23 " meaning percussion fuze, type 23. If any variation from the original design has been made, appropriate letters are added to the nomenclature ; the original serial number remains unaltered. The same procedure is applied when the fuze has been modified for use in projectiles other than those for which it was originally designed, or when redesigned for another weapon.

Examples : (*a*) A.Z.23.
(*b*) kl.A.Z.23 umg.
(*c*) le.Jgr.Z.23 n.A.
(*d*) A.Z.35K.
(*e*) Hbgr.Z.35K.

(*a*) Normal type (*see* above).
(*b*) Small and modified.
(*c*) New type for light infantry gun shell.
(*d*) Normal type.
(*e*) For use under ballistic cap.

Fuzes designed to provide for optional delay action are stamped with figures denoting the time of delay in seconds, e.g. (0.25). These figures are either included in the nomenclature or are stamped near the setting mechanism. In addition they are stamped with letters M V and O to indicate the setting for delay and non-delay action. The letters M and V are normally diametrically opposed, and the slot in the plug is brought into alignment when delay action is required.

Some fuzes are designed with alternative delay settings in which case the fact is included in the nomenclature stamped on the fuze, e.g. " A.Z.23 umg. m.2 V." This particular example has a safe position marked " + "

Percussion fuzes that are not numbered are designed for a specified equipment and the calibre is given in its nomenclature. Two examples are :—

3.7 cm Kpf. Z.Zerl.P (3.7 cm Flak).
5 cm ditto (5 cm Flak 41).

The letters " Zerl. P." signify a powder burning self-destroying element ; " Kpf.Z." means nose fuze.

Base percussion fuzes have the prefix letters " Bd.Z " stamped on them, and if they are of naval origin, the letter " C " with an oblique stroke preceding the fuze number. The nomenclature usually includes the calibre of the equipment and the nature of the projectile in which it is to be used.

Examples (Typical).

Bd Z.f 7.5 cm Pzgr = Base fuze for 7.5 cm A.P. projectile.
Bd Z.f 15 cm Gr. 19 Be = Base fuze for 15 cm anti-concrete shell, model 19.
Bd Z. C/38 = Naval origin base fuze type 38.

With these fuzes the M and V are usually stamped together as " MV " and the slot aligned with them to obtain delay action. If alternative delay is provided the stampings K/V (short delay) and G/V (long delay) will be found.

The " O " position on all types is for non-delay.

The presence of a delay unit in the older igniferous type of base fuze is not indicated by marking, but some of the oldest which are neither numbered nor designated by calibre are stamped with the letters " mV " on the underside indicating " with delay," i.e. a delay fitment not of the optional setting type.

Time and Time and Percussion Fuzes

German fuzes of this nature that have been captured are all of the mechanical type. (Fuzes of the combustion type that are in use have their origin in occupied countries.)

These fuzes do not belong to a numbered series. The number or numbers stamped on them refer to seconds of time.

Time fuzes are stamped with the abbreviated nomenclature " Zt.Z." or " Z.Z."

Time and Percussion fuzes are stamped " Dopp.Z."

In both types the nomenclature is followed by the letter " S " and an oblique stroke preceding the number(s) referred to above.

A single number denotes the maximum time interval from zero to which the fuze can be set. Two numbers, separated by an oblique stroke, indicate a fuze in which the maximum time of running has been altered in the design, the first figure referring to the actual time of running ; while two numbers separated by a hyphen denote the maximum and minimum time between which the fuze will function.

Examples :	Zt. Z. S/30	Fuze can be set to function from zero to 30 seconds.
	Dopp.Z. S/45-125 ...	Fuze will not function below 45 seconds. Maximum running time = 125 secs.

Additional stampings, such as Fg, nA, s, etc., are fully described in Appendix F.

Very little descriptive painting or stencilling is done on German fuzes, the only known marking of this nature being :—

K in black on the side of the fuze	Fuze safe for use in extreme cold.
Blue nose	Fuze for use in mountain guns. (This fuze is also stamped "Geb" in the fuze designation).
Yellow nose	Fuze manufactured from zinc.
Red band or ring	On the 3.7 cm Kpf.Z. Zerl. P., M.P., and M.A. fuzes to denote the self-destroying element does not function.

Gaines, Exploders and Smoke Boxes

German shell are normally issued fuzed to the troops in the field but loose gaines, exploders and smoke boxes may be met in ammunition dumps.

Gaines (Plate XVI)

Gaine bodies are made of steel, tinned brass or aluminium alloy, any marking being stencilled on the side in black or on a label stuck to the base. With the exception of some older types they are usually designated in a numbered series, sometimes with a letter and oblique stroke immediately in front of the number.

The following are typical of some of the markings that have been met :—

Zdlg. A	meaning	Gaine A.
Zdlg. B	,,	Gaine B.
Zdlg. C/98	,,	Gaine C/98.
gr.Zdlg. C/98	,,	Large gaine C/98.
Zdlg. C/98 Np	,,	Gaine C/98 filled with P.E.T.N. Wax.
kl. Zdlg.34 mV etc.	,,	Small Gaine No. 34 with delay, etc.

Exploders (and Burster Charges)

German H.E. shell filling designs do not as a rule include exploders. Any component of this nature that may be met will probably have its origin in occupied territory. However, there are small charges of German manufacture pressed into pellets, which may or may not be wrapped in waxed paper. These are used with or as opening charges in Smoke shell, etc.

The letters " Zdlg " or " Ldg Zd " are used in the designation of this component. For example: Kl.Ldg.Zd.K.Gr.Nb " is the marking on the burster charge in a 7.5 cm Kw.K (Tank) gun smoke shell.

The explosive in these components is coloured, e.g.

Cyclonite/Wax is coloured blue or bluish-green.

P.E.T.N./Wax is coloured pink.

Smoke Boxes (Plate XVI)

These are designated in a numbered series.

They are small cylindrical cartons, generally of waxed or varnished paper, with the nomenclature " Rauchentwickler " or " Rauchent W " followed by the series number printed on the side or on a label stuck to one end. The nomenclature includes, in an abbreviated form, the type of smoke composition " Phos " or " Phosphor," and the usual manufacturer's initials, etc. **Note.**—Red phosphorus is the only composition that has yet been met.

MORTAR AMMUNITION

Bombs

The system of marking on this store follows closely the lines enumerated under Gun Ammunition projectiles and a typical example is shown in figure 9. The stencilled markings are :—

(*a*) H.E. Code No.
(*b*) Without steel exploder container.
(*c*) Place, date and workmark of assembly.
(*d*) Weight class.
(*e*) Smoke Box No.
(*f*) Place, date and workmark of the H.E. filler.

Augmenting Charges

Flat rings normally enclosed in a stocking of silky material upon which is printed the following information :—

(*a*) The nomenclature of the bomb to which it belongs.
(*b*) Propellant weight, nature, shape, etc.
(*c*) Size or dimensions of the propellant.
(*d*) Place and year of manufacture of the propellant and delivery number.
(*e*) Initials of the factory where the charge was made up and the date.

Primary Cartridges

These are stamped on the base with the manufacturer's initials and year of manufacture and usually bear a stencilled numeral, the significance of which is not known.

MORTAR AMMUNITION.
Typical Marking.
Wgr ZT amh 43 153.
b.
a.
-13 oM Zn 10.7.43.w
c.
Stamping. 13 slz. 7.43
d.
e.
R 14
Stamping. A.dnr. 48
f. slz 5.7.43.L
Stamping. Amzl. 3b
AUGMENTING CHARGE.
Zn 7.5.43.
(0,2-67/42)
10 cm. Nb. W.35.
Ngl. Rg. P-12,5
1941/9/219.
dnr
a.
b.
c.
d.
e.
PACKAGE MARKING.
(White lettering.)
Munition
10cm. Nb.W.
dnf
5
42
PRIMARY CTGE.

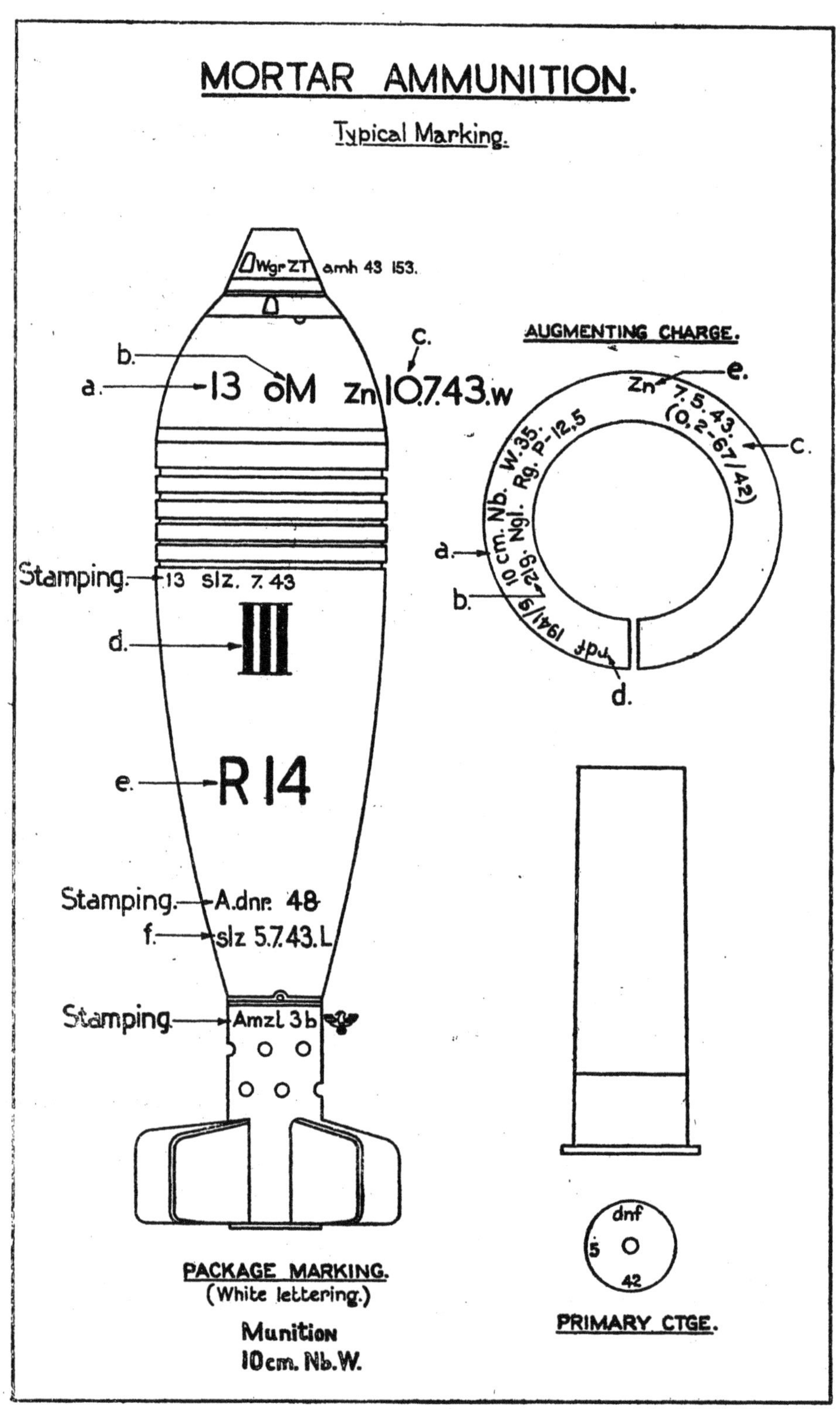

Fig. 9

APPENDIX A.

SMALL ARMS AMMUNITION—ABBREVIATIONS

Abbreviation	Full German Text	Translation
B-	Beobachtungsgeschoss	Incendiary-observing bullet
Br.	Brand	Incendiary.
E.	Eisenkern	Iron core.
El.	elektrische Abfeuerung	Electrically fired.
Ex.	Exerzierpatrone ...	Drill round.
f.	für	For.
—	gelb	Yellow.
G. or Gew.	Gewehr	Rifle.
Gesch.	Geschoss	Projectile, bullet.
Gl'spur	Glimmspur	Night tracer.
—	grün	Green.
H.	gehärtet	Hardened (often tungsten carbide).
i.L.	in Ladestreifen ...	In chargers.
K.	Kern (Stahlkern) ...	Steel core (A.P.).
K. or Karab	Karabiner	Carbine.
Kart	Kartusche	Cartridge.
kz.	kurz	Short.
l. or le	leicht	Light (weight).
lg.	lang	Long.
L'spur	Leuchtspur	Tracer.
M.-Gesch.	Minengeschoss	Indicates large H.E. capacity.
m	mit	With.
M.G.	Maschinengewehr ...	Machine gun.
M.P.	Maschinenpistole ...	Machine carbine.
n.A.	neue Art	New pattern.
Nahpatr.	Nahpatrone	Low velocity round.
o	ohne	Without.
Ph.	Phosphor	Phosphorus.
Patr	Patrone	Round (fixed cartridge).
Patrh.	Patronenhülse	Cartridge case.
Pist	Pistole	Pistol.
Pl	Platz	Blank.
Pz.B.	Panzerbüchse ...	A. tk. rifle.
—	rot	Red.
s.	schwer	Heavy.
S	Spitzgeschoss	Pointed bullet.
S *		(Indicates brass cartridge case).
S.E.	Sintereisen	Sintered iron.
Spr.	Spreng	H.E.
St.	Stahl	Steel.
Stahlh	Stahlhülse	Steel case.
—	Treib (patrone) ...	Propelling (cartridge).
u	und	And.
Üb	Übungs	Practice.
v.	verbessert	Improved.
vk.	verkurzt	Short (used for short burning tracer e.g. vk. L'spu
W.	Wärmeübertragung ...	Self-destruction by heat of tracer.
Z or Zerl	Zerlegung	Self-destruction.
Zdh	Zündhütchen	Percussion cap.
Z or Zdr	Zünder	Fuze.

APPENDIX B

NOMENCLATURE ABBREVIATIONS

The following abbreviations given with their German and English meanings are commonly found in the nomenclature of German ammunition. Further abbreviations are given under the separate sections dealing with the ammunition components. It should be noted that complex abbreviations are used by the Germans which are built up from the simple basic abbreviations and are therefore to be found listed under their separate parts.

German abbreviation	German meaning	English meaning
(a)	amerikanisch	American.
aA	alte Art	Old Model.
Al.	aluminium	Aluminium.
(b)	belgisch	Belgian.
Be	Beton	Concrete.
Cu	Kupfer	Copper.
(e)	englisch	English.
(E)	Eisenbahn	Railway mounting.
Ersst	Ersatzstück	Plug (e.g. fuse hole plug).
(f)	fazösisch	French.
f	für	For.
Fest	Festung	Fortification.
F.H.	Feldhaubitze	Field Howitzer.
F.K.	Feldkanone	Fd. Gun.
Fl	Flamme	Liquid incendiary.
Flak	Flugabwehrkanone	A.A. gun.
G	Geschütz	Low velocity gun.
Geb	Gebirgs	Mountain.
ger.	gerillt	Projectile grooved to control fragmentation.
gr	gross	Large.
Gr	Granate	Shell.
......Gr......Be......	Granate......Beton	Anti concrete shell.
......Gr......Hl......	Granate......Hohlladung ...	Hollow charge shell.
......Gr......Nb......	Granate......Nebel	Smoke shell.
Gr W.	Granatwerfer	Mortar.
(h)	holländisch	Dutch.
H	Haubitze	Howitzer.
Hb	Haube	Ballistic cap.
Hbgr	Haubegranate	Shell with ballistic cap.
H.T.	Haubitzetürm	Turret mounted howitzer.
(i)	italienisch	Italian.
(j)	jugoslavisch	Jugoslavian.
J.G. or I.G.	Infanteriegeschütz	Infantry gun.
K	Kanone	High velocity gun.
Kart	Kartusche	Q.F. separate Cartridge or Charge.
K.H.	Kanonenhaubitze	Gun howitzer.
K.K.	Kanonekasematte	Gun in Casemate.
kl	klein	Small.
KwK	Kampfwagenkanone ...	Tank gun.
kz	kurz	Short.
Kz	Kopfzünder	Nose fuse.
K.T	Kanonetürm	Turret mounted gun.
L/..........	Lang/..........	Length of shell or gun in calibres.
L or Laf	Lafette	Gun carriage.
Ldg	Ladung	Charge.
le	leicht	Light.
lg	lang	Long.
L.G.	Leichtes Geschütz	Light airborne gun (recoilless).

German Abbreviation	German meaning	English meaning
Lm	Leichtmetall	Light alloy.
L'spur	Leuchtspur	Tracer.
Lt Gs	Leucht Geschoss	Star shell.
M	Marine...	Naval.
m	mit	With.
Mgr.	Minengranate...	High capacity H.E. shell.
Mrs.	Mörser	Heavy howitzer.
Ms	Messing	Brass.
Mun	Munition	Ammunition.
(n)	norwegisch	Norwegian.
n.A.	neue Art	New type.
Nb or N	Nebel	Smoke.
Nbgr	Nebelgranate	Smoke shell.
NbW	Nebelwerfer	Smoke mortar or rocket projector.
n.F.	neue Fertigung	New method of production.
o	ohne	Without.
(ö)	österreichisch	Austrian.
(p)	polnisch	Polish.
Pak	Panzerabwehrkanone ... or Panzerjägerkanone	Anti-tank gun.
Patr.	Patrone	Q.F. fixed cartridge.
Pl Patr	Platzpatrone	Blank cartridge.
Pr	Pressstoff ... ... (except in the case of BoPr)	Moulded plastic.
Pz.B.	Panzerbüchse...	Anti-tank rifle.
Pzgr	Panzergranate	Anti-tank shot/shell.
(r)	russisch	Russian.
Rf.K.	Rückstossfreikanone... ...	Recoilless Gun.
R.Gr	Raketen Granate	Rocket assisted shell.
.....Rö Gr......Be...	Röchling Granate...Beton ...	Long anti-concrete shell.
	Rot	Shell with red smoke box.
	rot	This word, included in the designation of projectile, indicates the presence of a red band above the driving band. It is used to distinguish a projectile with a modified driving band where two otherwise identical types exist. Generally "rot" projectiles are fired from higher velocity equipments and the normal types from lower velocity equipments.
R Pz B	Raketenpanzerbüchse ...	Anti-tank rocket launcher.
R.W.	Raketenwerfer	Rocket projector.
s	schwer	Heavy.
Sf	Selbstfahrlafette	Self-propelled mounting.
SK	Schnelbladekanone	Q.F. gun.
Sn.	Zinn	Tin.
Sprgr	Sprenggranate	H.E. shell.
St	Stahl	Steel.
Stg	Stahlguss	Cast steel.
Stggr	Stahlgussgranate	Cast steel shell.
Stielgr	Stielgranate	Stick bomb.
Stu G.	Sturmgeschütz	S.P. assault gun (short barrel).
Stu K.	Sturmkanone	S.P. assault gun (long barrel).
(t)	tschechisch	Czech.
T.S.	Treibspiegel	Discarding sabot.
u	und	And.
umg	umgeändert	Modified.
W.G.	Wurfgerät	Rocket projector.
Wgr	Wurfgranate	Mortar or rocket bomb.
Wgr Patr	Wurfgranatpatrone ...	Mortar primary cartridge.
Wk	Wurfkörper	Rocket bomb.
Zdschr	Zündschraube	Primer.
Zldg	Zündladung	Gaine.
Zn	Zink	Zinc.

APPENDIX C

TYPES OF PROPELLANT

German Abbreviation	German meaning	English meaning
Digl or D	Diglykolpulver	Double base propellant of diethylene glycol dinitrate and nitrocellulose, stabilized with centralite and with potassium sulphate added to reduce flash.
Gu or GU	Gudolpulver	Diglycol powder with about 30 per cent. nitroguanidine added to reduce flash and as a stabilizer.
Ngl or NG	Nitroglyzerinpulver	Double base propellant of nitroglycerine and nitrocellulose stabilized with centralite, akardite or diphenylamine.
Nz or NZ	Nitrozellulosepulver	Single base nitrocellulose powder stabilized with diphenylamine and with sodium oxalate and potassium sulphate added to reduce flash.
NZ. Gew.P.	Nitrozellulosegewehrpulver ...	Small arms single base propellant of nitrocellulose stabilized with diphenylamine and including ethyl centralite and potassium sulphate.
Np. Gew. P	Nitropentagewehrpulver ...	Small arms double base propellant of P.E.T.N. and nitrocellulose, stabilized with diphenylamine and including ethyl centralite and potassium sulphate.
Nz. Man P	Nitrozellulosemanöverpulver.	Nitrocellulose powder with potassium nitrate incorporated to produce a porous powder used as an igniter.
Spr Schw P or Schw P	Sprengschwarzpulver ...	Gunpowder.

Propellant Granulation, etc.

Bl. P.	Blättchenpulver	Rectangular flakes.
Kr. R.	Kreutz Rohr	Central tube of propellant supporting charge in the case.
N P	Nudelpulver	Chopped Cord.
Pl P	Plättchenpulver	Multi-perforated discs.
Pol. P.	Polpulver	Powder produced without the use of a solvent.
R.P.	Röhrenpulver...	Tubular.
Rg P	Ringpulver	In the form of flat rings (Washers)
St P	Sternenpulver	Flat 6 pointed stars.
Stb	Stäbchen	Chopped Tube.
Str P	Streifenpulver	Strip.
W.P	Würfelpulver	Dice.

Other Markings

G	Pulvermasse G	Powder with a standard heat of explosion, in this case 690 calories.
Beildg	Beiladung	Igniter.
B.D.	Bleidraht	Decoppering foil/wire.
Trbldg	Treibladung	Propelling Charge.
V	Verbesserte Ladung ...	Adjusted Charge.

APPENDIX D

MISCELLANEOUS SHELL MARKINGS

Stencilling	Colour	Position (*see* Fig. 8)	German meaning	English meaning
A	White	C	Ausstoss	Ejector shell.
AB	Black	A or B or E	Ausstossbüchschen	Shell containing ejector boxes.
Al	Black	C	Sprengladung mit Aluminium-Griess	Granular aluminium flash producer incorporated in filling.
B or Buntr	Black	C	Buntrauch	Shell giving a red smoke burst.
Bl	White	C	Blindgeschoss	Inert filling.
blau	Black	C		Used together with Deut to indicate blue colour of smoke.
Bo or Bo Pr	Black	C	ausgebohrte Presstahlgranate	Forged steel shell with cavity bored out.
Br	White	C	Brandgranate	Incendiary shell.
Br. Schr.	Black	A.B.	Brand Schrapnelgeschoss	Incendiary Shrapnel Shell.
Deut	Black or White	C	Deutgeschoss	Indicator shell giving coloured smoke.
Cu	White	C D	Kupfer	Copper driving band.
Ei	Black	C	Einschiessgeschoss	Ranging shell.
Ex	Red	C	Exerziergeschoss	Drill round.
F	Black	C	Fern	Long range shell.
FES	White	C or D	Führung Sintereisen	Sintered iron driving band.
FEW	White	C or D	Führung Weicheisen	Soft iron driving band.
Hl or Hl/A etc.	Black	B	Hohlladung	Hollow charge anti-tank shell method of filling A, etc.
HK	Black or White		Hart Kern	Hard core.
K or Kt.	Black	C	Kartätsche	Case shot.
kg.	Black or White		Kilogram	Follows weight of fuzed shell in kilograms.
Kh	Black	E	Kammerhülse	Shell with central burster tube.
KPS	White		Kupfer Pressstahl	Bimetallic driving band.
Leucht	Black	C	Leuchtgeschoss	Star shell.
lg M	Black	A or B	lange Mundlochbüchse	Shell with lengthened gaine container.
Lm	Black	B	Duplex-Sprengkapsel (Lm)	Shell fuzed with a combined cap and gaine in aluminium.
Lo	Black	A	lose Sprengstoff-körpern	H.E. shell consisting of separate explosive bodies.
LS	White	B or C	Leuchtsatzsprengladung	Illuminator shell.

Stencilling	Colour	Position (*see* Fig. 8)	German meaning	English meaning
m R	Black	B	mit Rauchentwickler	Shell containing smoke box.
Nb or N	White	C	Nebelgeschoss	Smoke shell.
O	Black	C	ohne Füllung	Marking of 20 mm. shot.
o M	Black	B	ohne Mundlochbüchse	Shell without gaine container.
o R	Black	C	ohne Rauchentwickler	Shell without smoke box.
P.G. or P	Black	C	Perlitgussstahl	Shell of cast steel in the pearlite condition.
P͡h	Black	C	Phosphor	Phosphorus incendiary filling.
R 11 etc.	Black	C	Rauchentwickler Nr. 11	Shell containing smoke box No. 11, etc.
Rot	Black	C		H.E. shell giving a red smoke burst.
R S	Black	C	Reizstoff	Shell containing irritant filling.
Spr Br	Black	C	Spreng Brand	H.E./incendiary filling.
Stg	Black	C	Stahlgussgranate	Cast steel shell.
ohne Al	White	C	ohne Aluminium	Aluminium, normally included in the filling, has been omitted.
	Black	D		Shell with modified driving band for firing from Czech guns.
Tp	Red or Black	C	Tropen	Shell filled suitable for tropical use.
U	Black	C	Unterrichtsgeschoss	Instructional shell.
Üb	White	C	Übungsgeschoss	Practice shell filled gunpowder.
Üb Al	White	C	Übungsgeschoss mit Aluminium Sprengladung	Practice shell filled gunpowder and aluminium powder.
Üb B	White	C	Übungsgeschoss B	Practice shell filled H.E. with an SO_3 smoke box.
Üb T	White	C	Übungsgeschoss T	Practice shell filled H.E. with Tetrachlornaphthalene incorporated.
Üb W	White	C	Übungsgeschoss Weiss	Practice shell giving white smoke burst.
Üb R	White	C	Übungsgeschoss Rot	Practice shell giving red smoke burst.
Üb S	White	C	Übungsgeschoss Schwarz	Practice shell giving black smoke burst.
vk	Black	B	verkürzt	Shortened.
Vp	Black	C	Verpackungsgeschoss	Dummy round for packing, etc., trials.
W	White	C	Weichkern	Soft iron core.
wKh	White	C	weite Kammerhülse	Wide central burster tube.
Z B	Black	E	Zwischenboden	Diaphragm shell.

APPENDIX E

CODE NUMBERS DENOTING SHELL FILLING

These will be found stencilled in black in arabic numerals in areas B or C. (*see* Fig. 8).

It should be noted that the proportions given are average since wide variations are encountered.

Number	Meaning
1	T.N.T. pressed in blocks in cardboard container packed with magnesium putty.
1A	T.N.T. pressed in blocks in cardboard container packed with paper.
1B	T.N.T. pressed in blocks in cardboard container packed with montan wax in a metal container.
2	Picric acid pressed in blocks in cardboard containers.
3	P.E.T.N. pressed in blocks.
4	T.N.T. loose in paper carton.
5	Picric acid loose in paper carton.
6	T.N.T./Wax (95/5) in blocks in cardboard carton.
7	T.N.T. pressed.
8	T.N.T. poured.
9	
10	T.N.T. + T.N.T./Wax (95/5) + T.N.T./Wax (90/10) pressed in blocks in cardboard cartons.
11	T.N.T. + T.N.T./Wax (90/10) + T.N.T./Wax (85/15) + T.N.T./Wax (80/20) pressed in blocks in cardboard cartons.
12	T.N.T. + T.N.T./Wax (95/5) + Cyclonite/Wax (90/10) pressed in blocks in cardboard cartons.
13	Amatol 40/60 poured.
13A	Amatol 50/50.
14	T.N.T. poured.
15	T.N.T./Aluminium powder (90/10) poured.
16	T.N.T. poured in aluminium container + PETN/Wax (90/10) as exploder.
17	T.N.T./Aluminium powder (90/10) poured + PETN/Wax (90/10) as exploder.
17A	Matrix of Dinitroanisole/Ammonium nitrate/Cyclonite (54/32/14) with biscuit of Ammonium nitrate/Calcium Nitrate/Cyclonite/Pentaerythritol/combined water (46/21/20/9/4).
18	T.N.T./Cyclonite wax (80/19/1) pressed in blocks in cardboard cartons.
19	T.N.T./Ammonium nitrate/Aluminium powder (55/35/10) poured.
20	Ethylenediamine Dinitrate/Ammonium nitrate/Aluminium powder (53.5/1.5/45).
21	Amatol 40/60 with core of pressed T.N.T. pellets.
22	
23	
24	Picric acid poured.
25	
26	
27	T.N.T. + T.N.T./Wax (90/10) pressed in blocks in cardboard cartons.
28	T.N.T./Wax (90/10) + PETN/Wax (90/10) pressed in blocks in aluminium container.
29	T.N.T./Wax (90/10) + T.N.T. crystallized + T.N.T./Wax/Potassium Chloride (63/7/30) + T.N.T./Wax/Potassium Chloride (45/5/50) + Potassium Chloride in blocks in cardboard carton.
30	T.N.T. + T.N.T./Wax (95/5) pressed in blocks in cardboard carton.
31	
32	PETN/Wax (90/10) pressed in waxed paper wrapping.
33	PETN/Wax (85/15) pressed in waxed paper wrapping.
34	PETN/Wax (70/30) pressed in waxed paper wrapping.
35	
36	PETN/Wax (60/40) pressed in blocks in waxed paper wrapping.
37	
38	PETN/Wax (35/65) pressed in blocks in waxed paper wrapping.
39	
40	
41	
42	

Number	Meaning
43	Plastic PETN.
44	
45	PETN/Wax/Cyclonite (35/15/50)
46	
47	
48	
49	
50	
51	
52	Dinitrobenzene/Ammonium nitrate/Cyclonite (50/35/15) poured.
52+	Matrix of Dinitrobenzene/Ammonium nitrate/Cyclonite (47/38/15) with biscuit of Ammonium nitrate/Calcium nitrate/Cyclonite/Pentaerythritol/combined water (46/21/20/9/4).
52A	Matrix of Dinitrobenzene/Ammonium nitrate/Cyclonite (50/35/15) with biscuit of Ammonium nitrate/Calcium nitrate/Cyclonite/Pentaerythritol/combined water (46/21/20/9/4).
52A+	Matrix of Dinitrobenzene/Ammonium nitrate/Cyclonite (53/30/17) with biscuit of Ammonium nitrate/Calcium nitrate/Cyclonite/Pentaerythritol/combined water (46/21/20/9/4).
53	
54	
55	
56	Donarit (Nitroglycerine/Ammonium nitrate/TNT/Rye flour (4/80/12/4)).
57	Monachit (Ammonium nitrate/Alkali nitrate/TNT/AlkaliChloride/Collodion cotton/Charcoal (64/3/14/17/1/1)).
58	
59	
60	Trinitrochlorobenzene.
61	Trinitrochlorobenzene poured.
62	
63	
64	Trinitrochlorobenzene/Ammonium nitrate (60/40) pressed.
65	
66	PETN/Wax (50/50).
67	
68	
69	
70	Trinitrobenzene pressed.
71	
72	
73	
74	
75	
76	
77	
78	
79	
80	
81	
82	
83	Ethylenediamine dinitrate pressed.
84	Ethylenediamine dinitrate/Ammonium nitrate (55/45).
85	
86	Ethylenediamine dinitrate/Cyclonite/Wax (46/18/36) pressed in blocks wrapped in waxed paper in aluminium container.
87	
88	
89	
90	Cyclonite pressed.
91	Cyclonite/Wax (95/5) pressed in blocks wrapped in waxed paper.
92	Cyclonite/Wax (90/10) pressed in blocks wrapped in waxed paper.
93	
94	
95	Cyclonite/TNT/Wax (60/38/2) pressed in blocks wrapped in waxed paper.
96	
97	Cyclonite/TNT (60/40) Cast.
98	
99	
100	

Number	Meaning
101	TNT/Wax (85/15).
102	Amatol (40/60) + Wax.
103	
104	Cyclonite.
105	TNT/Cyclonite/Aluminium powder (70/15/15) poured.
106	TNT/Cyclonite/Aluminium powder (50/25/25) poured.
107	
108	
109	Cyclonite/Aluminium powder/Wax (70/25/5) pressed.
110	Ammonium nitrate/Naphthalene/Aluminium powder/Wood Meal (90/5/2.5/2.5).
111	Ammonium nitrate/Carbon (96/4).
112	Amatol (20/80).
113	Ammonium Nitrate/TNT/Aluminium powder (70/20/10).

Additional abbreviations used in the description of the filling.

Abbreviation	Meaning
Fp 02	T.N.T.
Fp 40/60	Amatol 60/40.
Fp 5, etc.	T.N.T. + 5, etc., per cent. wax.
Grf 88	Picric acid.
Np 5 etc.	PETN + 5, etc., per cent. wax.
H 5 etc.	Cyclonite + 5, etc., per cent. wax.

Additional arabic numerals found in the stencilling on projectile bodies.

Stencilling	Colour	Position (*see* Fig. 8)	Meaning
12.2	White	A or E and Base	Calibre of shell 12.2 cm.
15.2	White	A or E and Base	Calibre of shell 15.2 cm.
15.5	White	A or E and Base	Calibre of shell 15.5 cm.
35	Black	C	F H Gr 35.
36	Black	B or C	Gaine No. 36.
38	Black or White	C, D or Base	F H Gr 38.
39	Black or White	C, D or Base	F H Gr 39.
40 Nb	White	C and Base	F H Gr 40 Nb.
41	White	C and Base	F H Gr 41.
Arabic Numerals	Black	Rear of C or D	Weight of shell in kg. Used with foreign shell.
1	White	A. Below white tip on the cap of 8.8 cm APCBC shell	Pzgr. Patr. 39-1.

APPENDIX F

The following abbreviations are used in the stampings on fuzes. In addition some abbreviations given in Appendix B may also be employed.

German abbreviation	German meaning	English meaning
A.Z.	Aufschlagzünder	Percussion fuze.
B.Z.	Brennzünder	Combustion time fuze.
Bd.Z.	Bodenzünder	Base fuze.
C/......		Model ... naval type fuze.
Dopp Z.	Doppelzünder	Time and percussion fuze.
E Kzdr or E Kz	Empfindlicher Kopfzünder	Sensitive percussion fuze.
E R Z	Electrischer Randdüsenzünder	Electric rocket igniter.
Fl, Fg	Fliehgewichtsantrieb	Operated by centrifugal weights.
Hbgr	Haubengranatzünder	Nose fuze for use under a ballistic cap.
Jgr Z.	Jnfanteriegranatzünder	Percussion fuze for use with inf. gun shells.
.....K.....	Kanone	Included in description of fuze for use with high velocity guns.
kl AZ	kleiner Aufschlagzünder	Percussion fuze for fitting to shell with small fuze hole.
Kpf Z or KZ	Kopfzünder	Nose fuze.
Lg Zdr	Leuchtgeschosszünder	Time fuze for use with star shell.
m K or m K+ or m Ko+	mit Klappensicherung	With shutter safety device.
m V or m	mit Verzögerung	With delay.
o AZ	ohne Aufschlagzündung	Without percussion element.
o V or o	ohne Verzögerung	Without delay.
s	schwer	Fuze for use in guns with high shell acceleration.
..... S/30		Time fuze with maximum running time of 30 secs.
..... S/90/45		Time fuze with maximum running time of 45 secs. modified to 90 secs.
..... S/45-125		Time fuze with no setting possible below 45 secs. and with maximum running time of 125 secs.
T	Trolitul	Plastic fuze body.
Vz		Model ... for Czech fuzes.
v	vereinfacht	Simplified.
V	Verzögerung	Delay.
Wgr Z	Wurfgranatzünder	Mortar bomb fuze.
Zerl	Zerleger	Self-destruction element.
Zerl Fg	Fliehgewichtszerleger	Centrifugally operated self-destruction element.
Zerl P	Pulverzerleger	Powder burning self-destruction element.
Zt Z.	Zeitzünder	Time Fuze.

APPENDIX G

ABBREVIATIONS AND NOMENCLATURE OF GERMAN GRENADES

The principal types of grenade in use by the Germans are :—

(*a*) Hand Grenades.

(*b*) Rifle Grenades.

(*c*) Kampfpistole Grenades (2.7 cm.) (Battle Pistol).

(*d*) Leuchtpistole Grenades (2.7 cm.) (Signal Pistol).

When compared with the standardized system of abbreviations used for Gun ammunition, the recognition of grenades by designation alone is somewhat difficult. This is largely owing to the fact that many terms are loosely applied and that some nomenclatures are not fully descriptive of the item.

The following terms, either in the abbreviated form or in full, are used on labels and as ammunition markings.

German abbreviation	German word	English meaning
B.Gr.	Blendgranate	Bursting smoke grenade (A.tk.)—Rifle Discharger type.
B.K.	Blendkörper	As above—Hand thrown type.
B.Z.	Brennzünder	Friction pull igniter.
Deut. Z.	Deutpatrone für Kampfpistole	Coloured smoke cartridge for the Battle Pistol.
Eihgr. or Eihdgr.	Eihandgranate	Egg-shaped hand grenade.
F.Leucht Z	Fallschirmleuchtpatrone für Kampfpistole	Illuminating Star on Parachute Cartridge for the Battle Pistol.
F. Sig. Z	Fallschirmsignalpatrone (grün) fur Kampfpistole	Green Signal Star on Parachute Cartridge for the Battle Pistol.
F.S.Lt.Gr.	Fallschirmleuchtgranate	Illuminating (with parachute) grenade.
G / Gw. / Gew.	Gewehr	Rifle.
G.Kart	Gewehr Kartusche	Propelling cartridge for rifle grenade (Blank).
G. Treibpatr.	Gewehr Treibpatrone	As above (Bulletted blank).
gr.	gross	Large.
Gr.B. 39	Granatbüchse 39	7.9/13 mm. A.tk. rifle modified to fire grenades.
Hgr. or Hdgr.	Handgranate	Hand grenade.
Hohl. or Hl	Hohlladung	Hollow charge.
K (generally in full)	Kampfpistole	Battle Pistol (with rifled barrel) (2.7 cm.)
L.P.	Leuchtpistole	Smooth-bore signal pistol (2.7 cm.).
Nachr.Z.	Nachrichtpatrone für Kampfpistole	Message cartridge for the Battle Pistol.
Nb.	Nebel	Smoke.
Nb. Gr.	Nebelgranate	Smoke grenade.
Nebel Z.	Nebelpatrone für Kampfpistole	Smoke cartridge for the Battle Pistol

German abbreviation	German meaning	English meaning
Propgr.	Propagandagranate	Grenade or bomb filled with propaganda leaflets.
P.W.M.	Panzerwurfmine	Hollow charge hand grenade.
Pzgr.* or Pz.Gr. or Pz.Wk.	Panzergranate	Hollow charge grenade.
—	Schiessbecher	Rifled cup discharger (3.0 cm.).
Spl.Rg.	Splittering	Fragmentation sleeve for hand grenade.
Spl.M.	Splittermantel	Fragmentation jacket for egg grenade.
Spr.K.Nr.8	Sprengkapseln Nr. 8	Detonators, No. 8.
Spr.K.Nr.8 (Al)	do. (Al)	do. (Aluminium).
Spreng.Z.	Sprengpatrone für Kampfpistole	H.E. Cartridge for the Battle Pistol.
Sprgr.Patr.K. or Sprgr. Patr. K.P.	Sprenggranatpatrone für Kampfpistole	As above (different model).
S.S.	Schutzstaffel	Nazi S.S. troops.
Stielhgr. or Stielhdgr.	Stielhandgranate	Stick hand grenade.
umg	umgeändert	Converted, modified.
Wgr.Patr.	Wurfgranatpatrone	H.E. signal pistol grenade (also primary cartridge for mortar bomb).
Wk.	Wurfkörper	H.E. signal pistol grenade (also Rocket bomb).
Zdlg. N 4	Zündladung N 4	Ignition tube No. 4.
Zdschn Anz	Zündschnuranzünder	Safety fuze igniter.
Zt Z	Zeit Zünder	Time fuze.

* When applied to Artillery ammunition Panzergranate means A.P. shot or shell.

TYPICAL GERMAN S.A.A. PACKAGE LABELS.

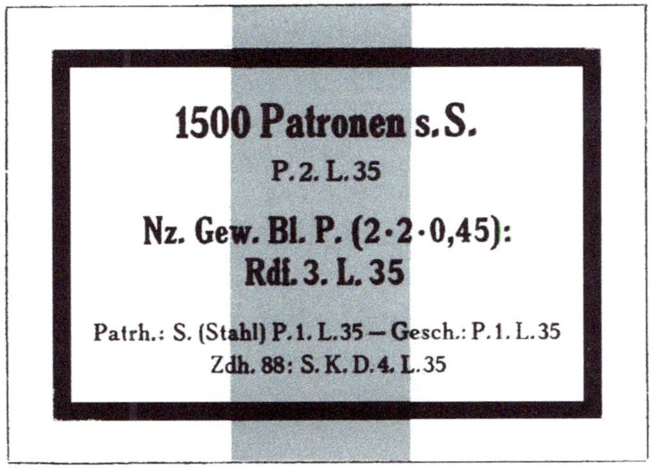

1500 Patronen s.S.
P.2.L.35
Nz. Gew. Bl. P. (2·2·0,45):
Rdf. 3. L. 35
Patrh.: S. (Stahl) P.1. L.35 – Gesch.: P.1. L.35
Zdh. 88: S.K.D.4. L.35

(7·92 BALL)

1500 Patronen S.m.K. L'spur (grünrot)
P.1. L.35
Nz. Gew. Bl. P. (2·2·0,45): Rdf. 17. L.35
Patrh.: S.*P.69.5.L.35 – Gesch.: P.69.20.L.35
Geschoßteile: P.69 – Satz: R. W. S.
Zdh. 88: S. K. D.20. L.35
Trocken aufbewahren! Gegen Stoß und Fall schützen!

(7·92 AP/T.)

1500 Patronen S.m.K.
P.2. L.35
Nz. Gew. Bl. P. (2·2·0,45):
Rdf. 128 1.L.35
Patrh.: S.* P.57. L.35 – Gesch.: P.77. L.35
Geschoßteile: P. – Zdh.88: S. K. D.58. L.35

(7·92 A.P.)

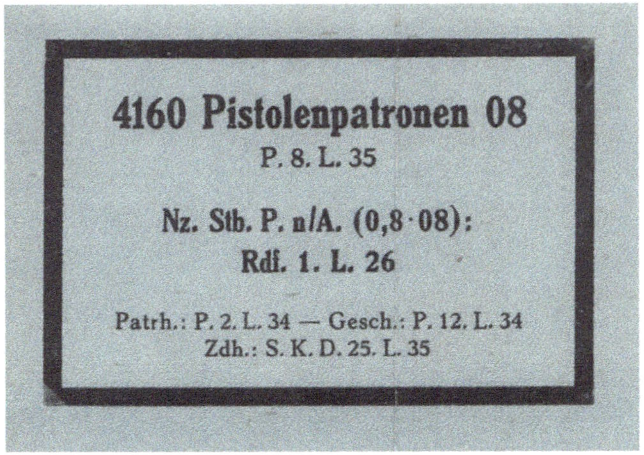

4160 Pistolenpatronen 08
P. 8. L. 35
Nz. Stb. P. n/A. (0,8·08):
Rdf. 1. L. 26
Patrh.: P. 2. L. 34 – Gesch.: P. 12. L. 34
Zdh.: S. K. D. 25. L. 35

9 mm. (PARABELLUM) PISTOL AMMUNITION.

The above labels are reduced in size for smaller packages.

Malby & Sons, Lith.

L.2525. 48239. 2490. 10,000, 6/45.

L2525.

GERMAN 7·92 mm. S.A.A.

This amn. is interchangeable with British Besa and vice versa.

(a) GROUND USE.

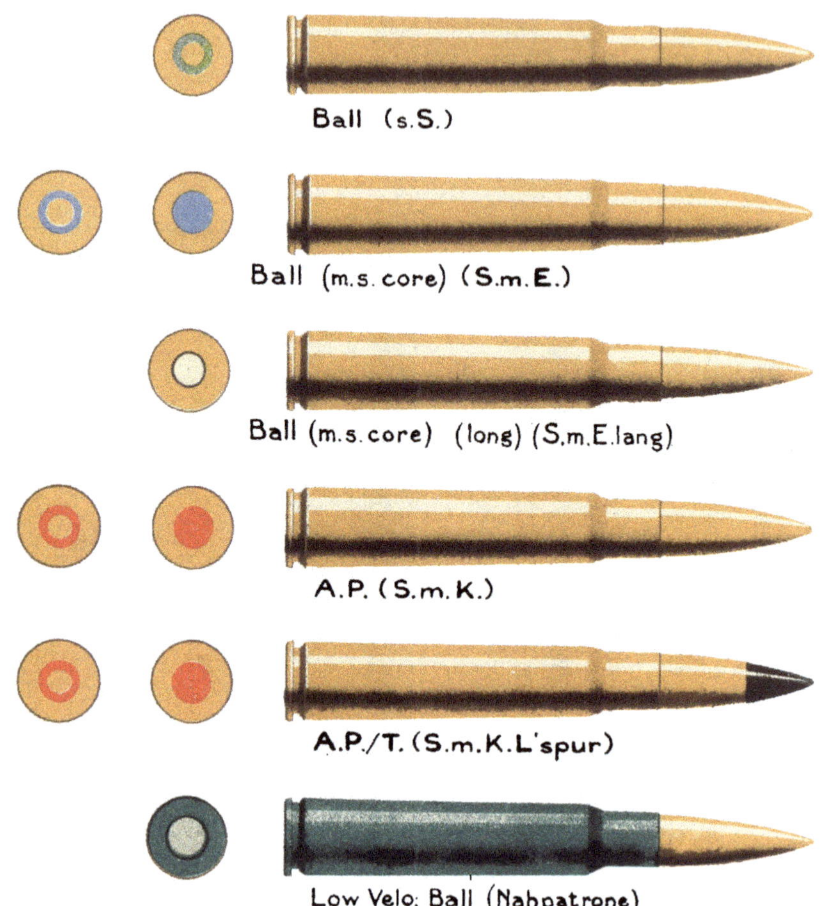

(b) MISCELLANEOUS

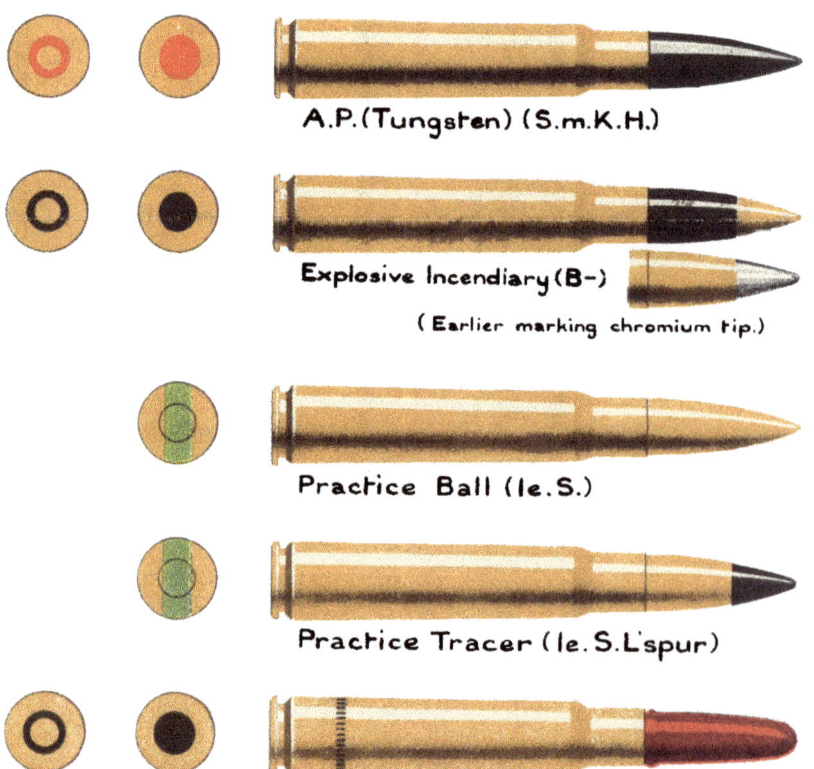

Malby & Sons, Lith.

PLATE II

(c) FOR USE IN AIRCRAFT GUNS ONLY.

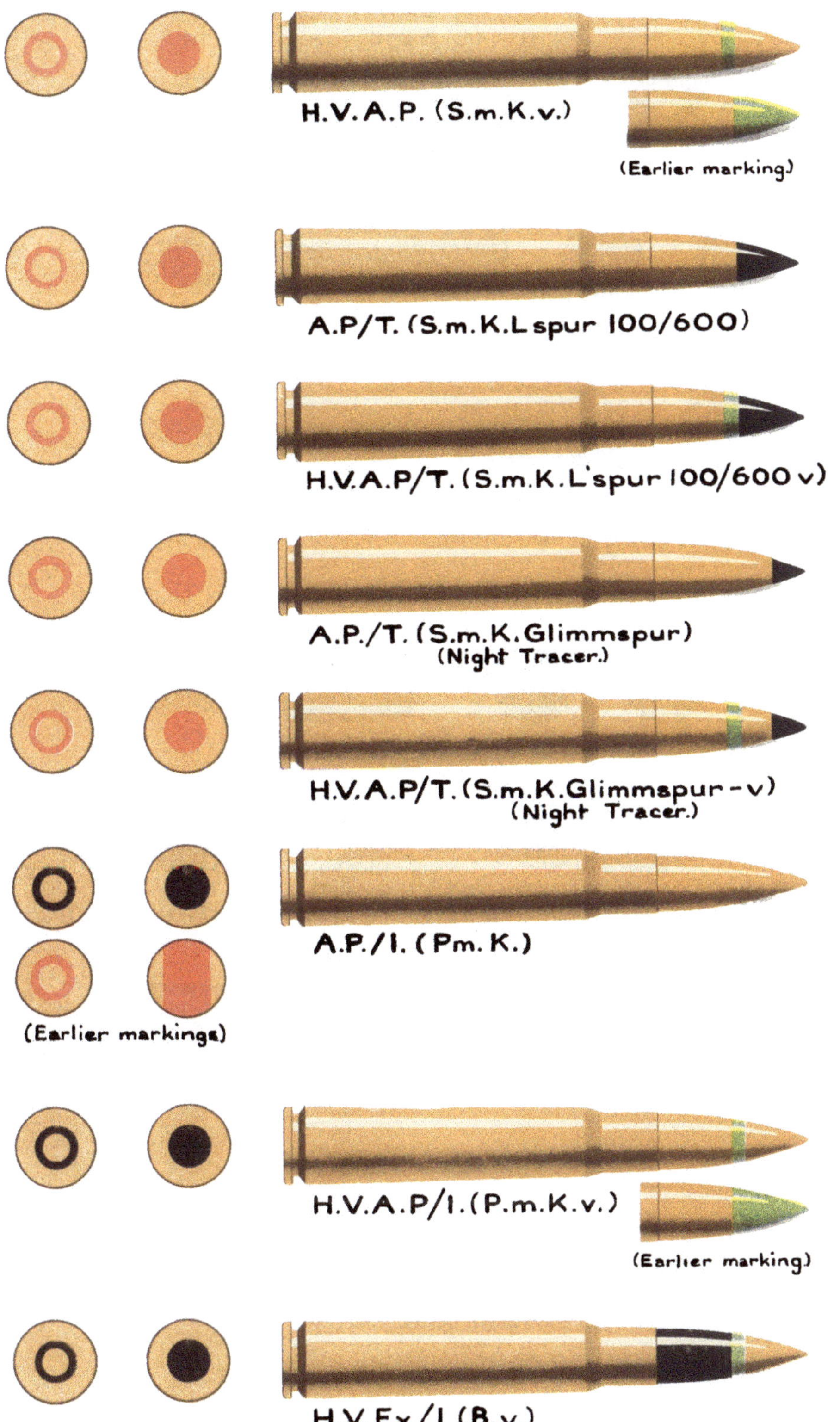

PLATE III

L 2525

Malby & Sons, Lith

(d) RIFLE GRENADE PROPELLING CARTRIDGES

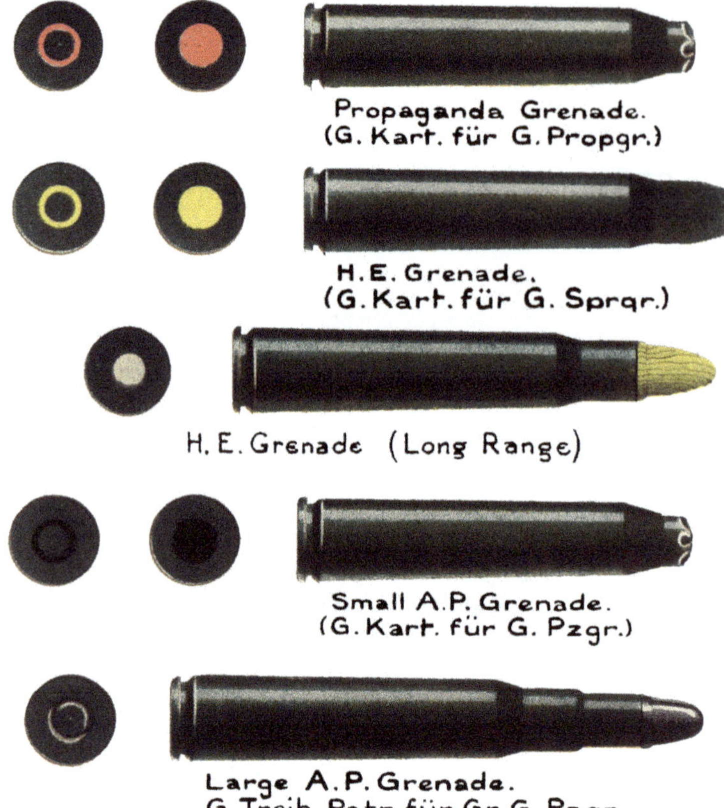

(e) PISTOL and MACHINE CARBINE AMMUNITION

Propelling Cartridge for Granatbüchse 39.

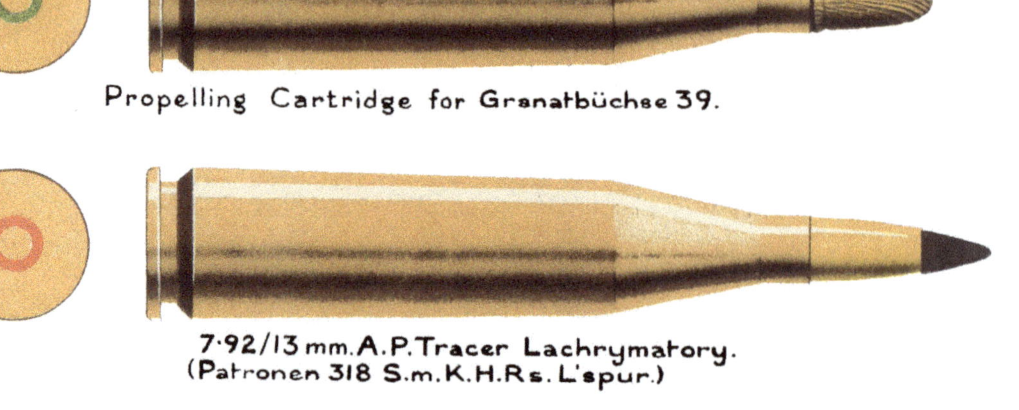

Malby & Sons, Lith.

PLATE IV

GERMAN 2cm. SOLOTHURN SHELLS.

Malby & Sons, Lith.

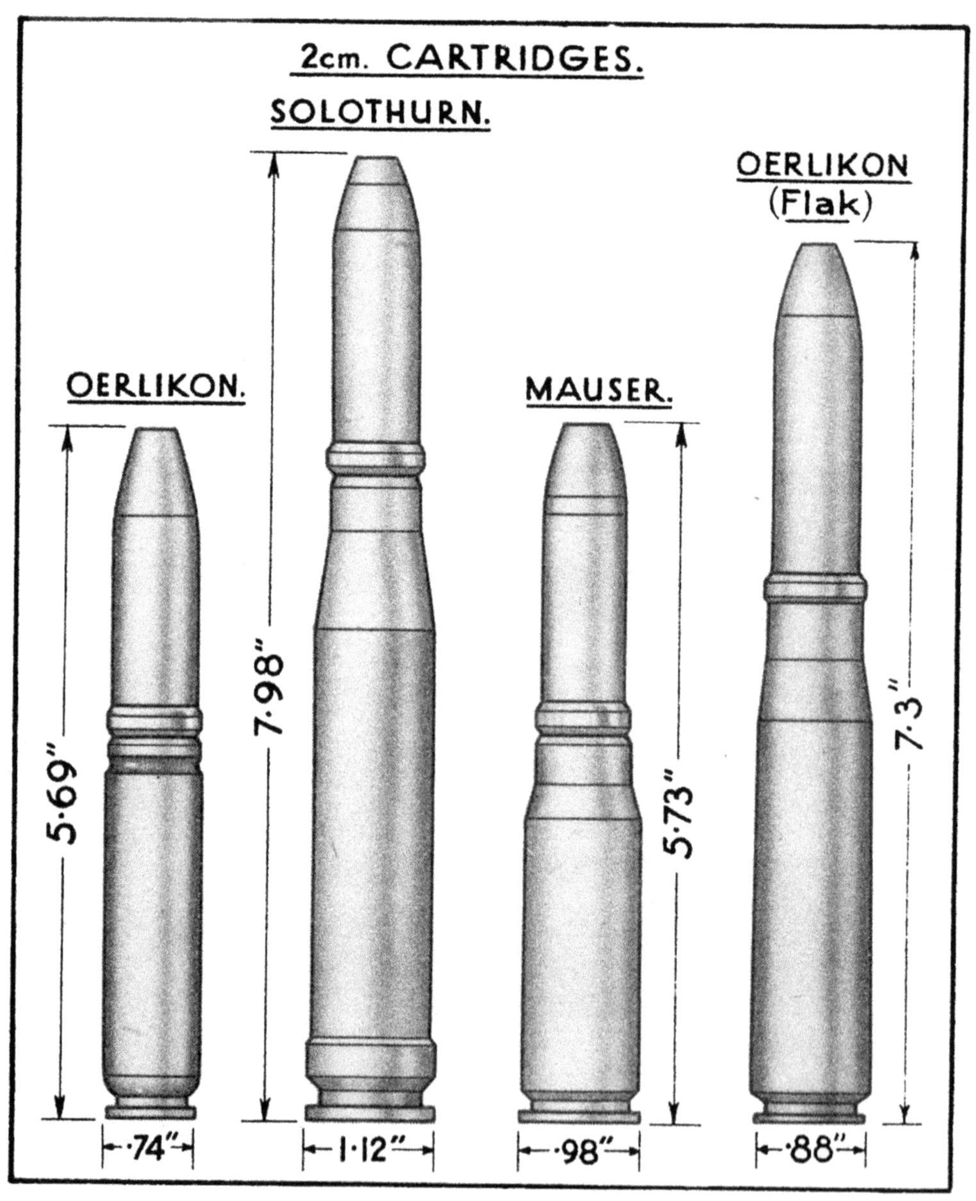

Malby & Sons, Lith.

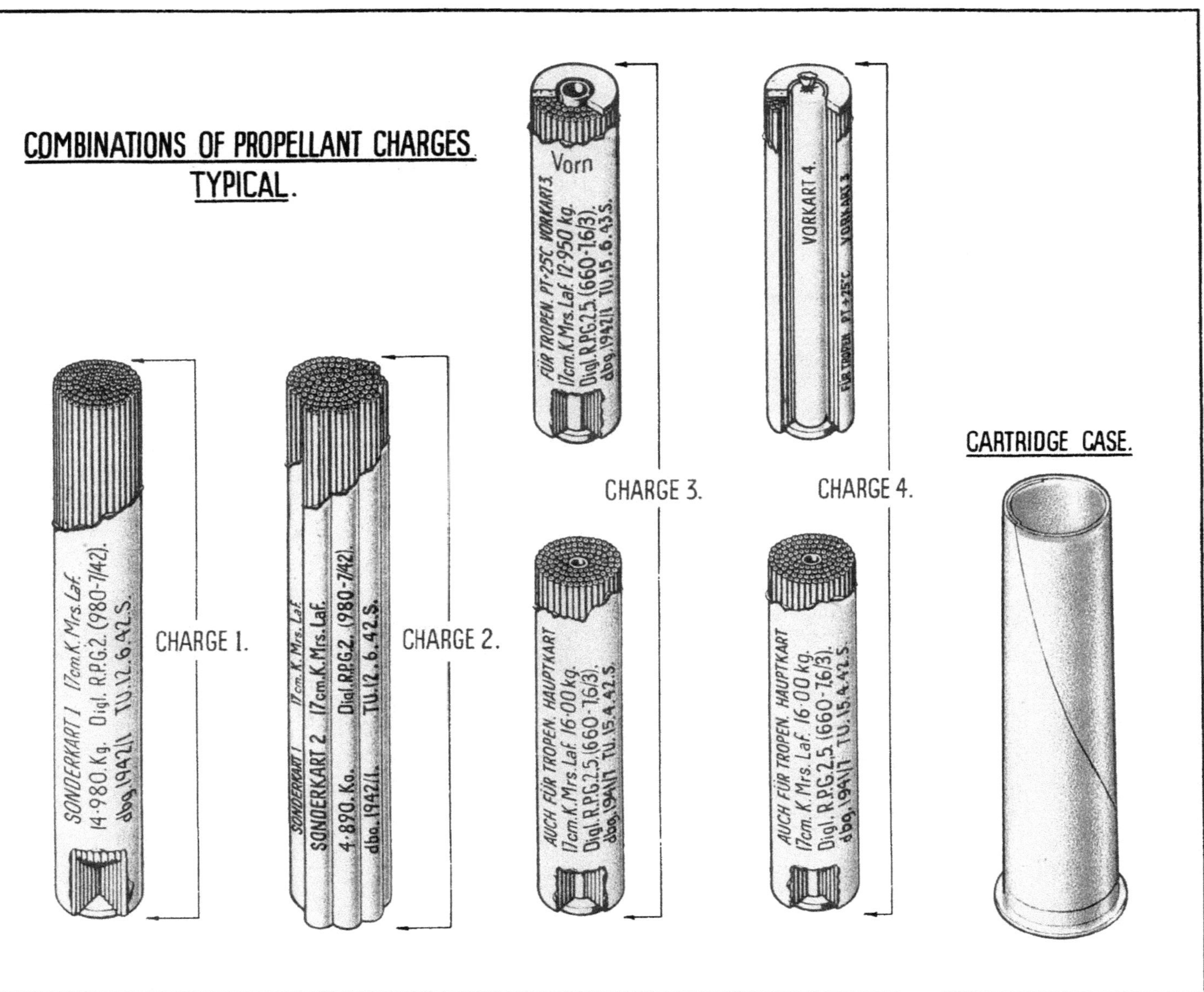

L.2525. Malby & Sons, Lith.

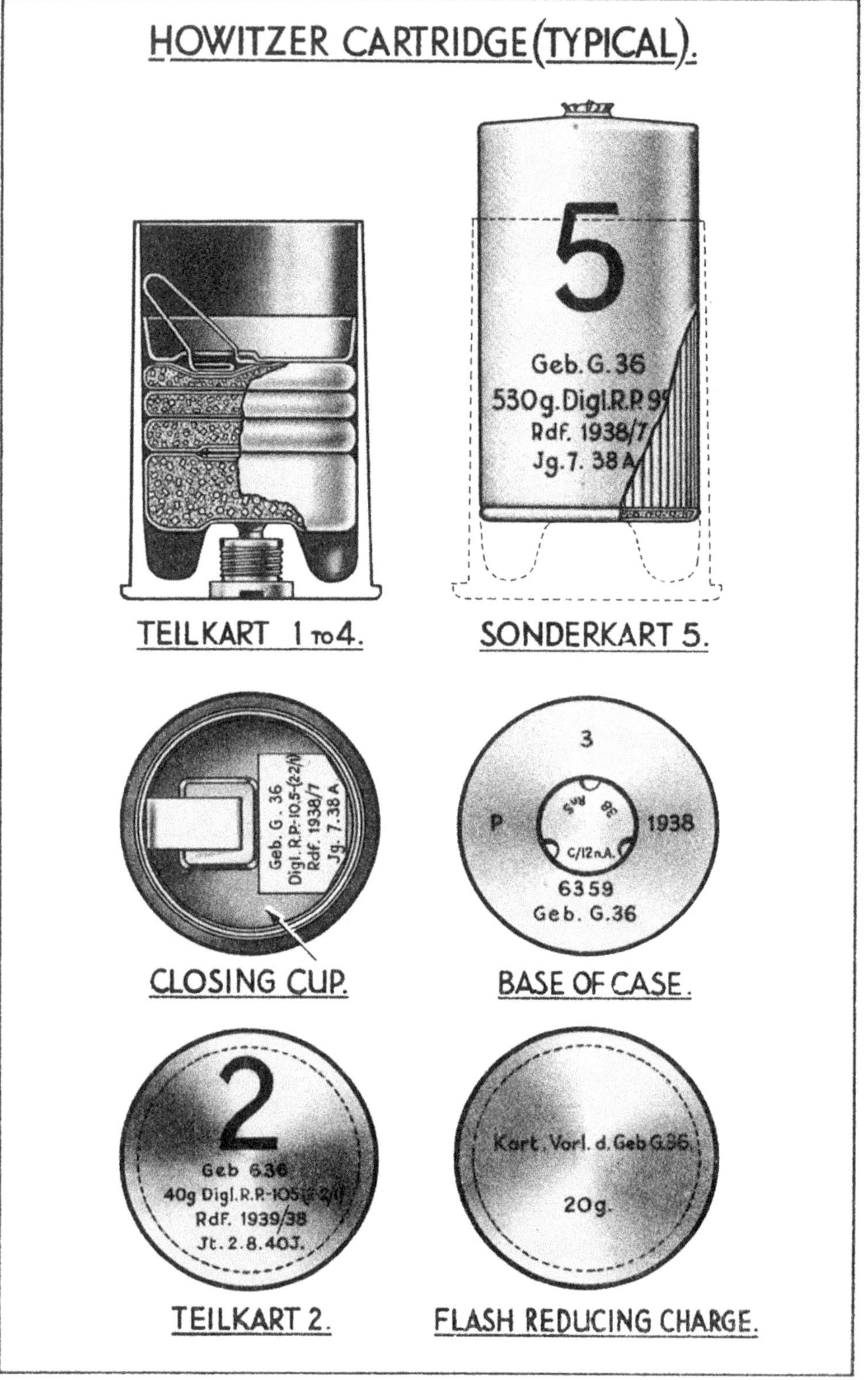

L.2525.

Malby & Sons, Lith.

TYPICAL Q.F. CARTRIDGES (LIGHT ANTI-TANK GUNS).

L. 2525.

H.E. SHELL.

Sprgr. Patr.

A.P. SHELL.
(FILLED H.E.)

Pzgr. Patr.

A.P. SHOT.
(WITH T.C. CORE)

Pzgr. Patr. 40 m HK.

PLATE IX

Malby & Sons, Lith.

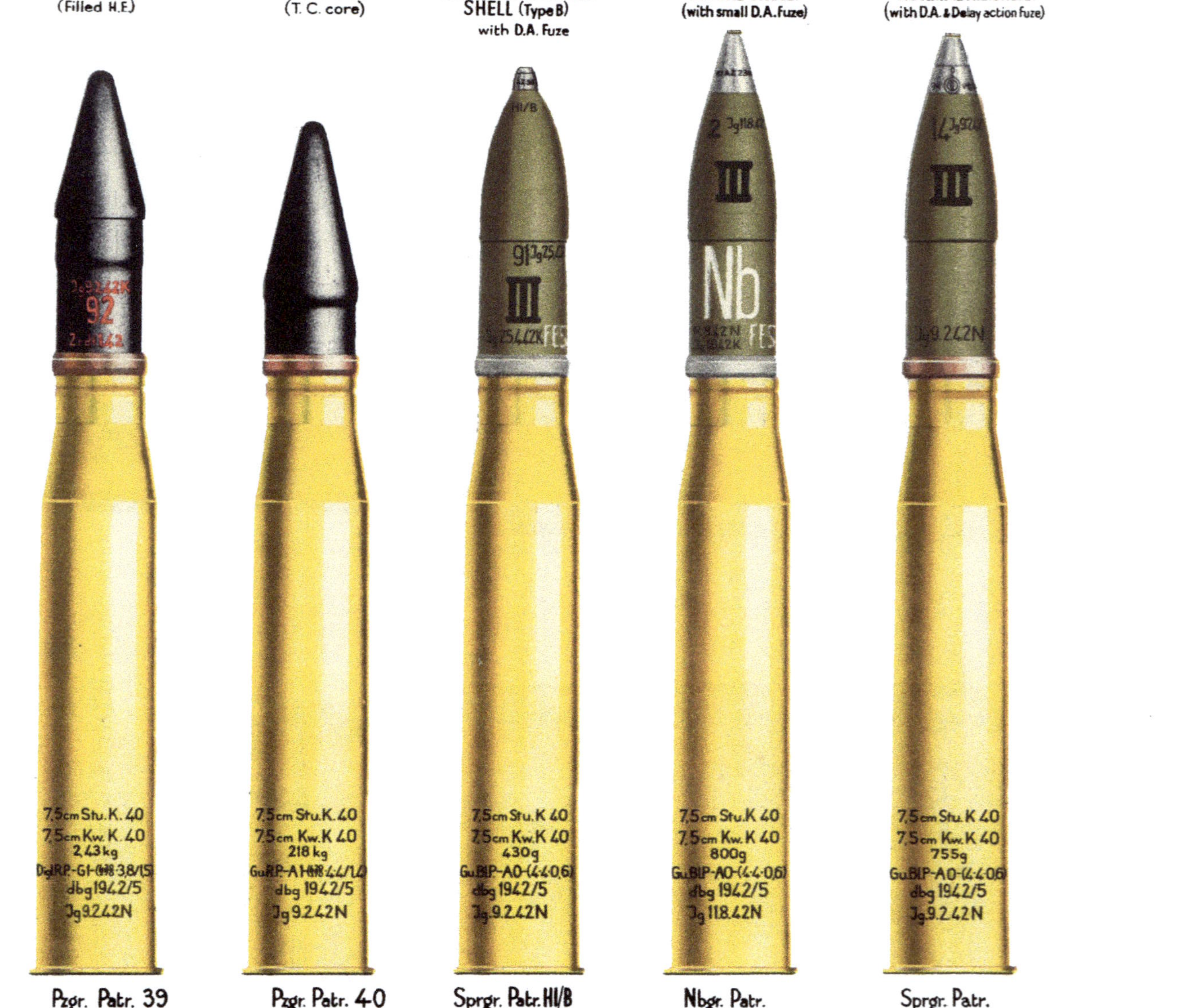

L. 2525.

Malby & Sons, Lith.

TYPICAL Q.F. CARTRIDGES (HEAVY A.A./A.Tk. GUNS)

H.E. SHELL
(with time fuze)

Sprgr. Patr.

A.P.C.B.C. SHELL

Pzgr. Patr. 39

Malby & Sons. Lith.

PLATE XI

L. 2525.

TYPICAL Q.F. CARTRIDGES (HEAVY ANTI-TANK GUNS)

H.E. SHELL

Sprgr. Patr. 43

A.P.C.B.C. SHELL

Pzgr. Patr. 39/43

A.P. SHOT
T. C. CORE

Pzgr. Patr. 40/43

HOLLOW CHARGE SHELL

Gr. Patr. 39/43 HL

PLATE XII

Malby & Sons, Lith.

PLATE XIII

TYPICAL SHELL (SHOWING THEIR MARKINGS.)

SMOKE SHELL.

BURSTING TYPE.
(WITH SMALL DIRECT ACTION PERCUSSION FUZE.)

(F.H.Gr 38Nb)

BASE EJECTION TYPE.
(WITH MECH: T&P FUZE)

(F.H.Gr 40 Nb)

HIGH EXPLOSIVE SHELL.

INCORP: COLOURED SMOKE.
(WITH MECH: T&P FUZE.)

F.H.Gr....Deut

NORMAL.
(WITH MECH: T&P FUZE.)

(F.H.Gr 38 Stg.)

HOLLOW CHARGE
(WITH D.A. FUZE.)

Gr...H1/C

ARMOUR PIERCING WITH BALLISTIC CAP.
(WITH BASE FUZE & TRACER.)

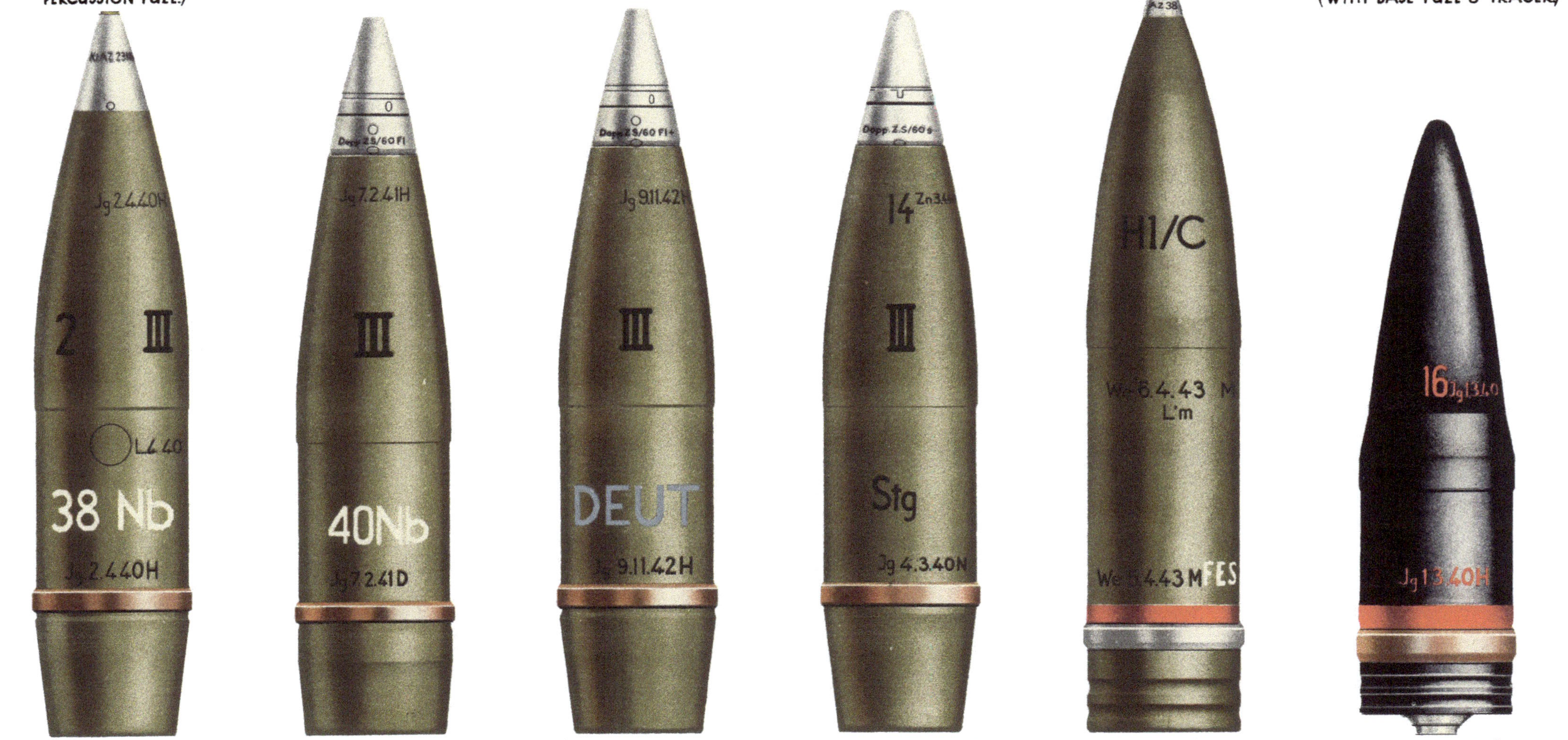

(Pzgr 39)

L.2525.

Malby & Sons. Lith

L. 2525.

HEAVY & MEDIUM GUN OR HOW. SHELL.

H.E. SHELL.

WITH DELAY ACTION PERCUSSION FUZE AND Nº 9 SMOKE BOX.

(K.Gr.18. mit A Z 23v (0,15))

ANTI-CONCRETE SHELL.

THIS SHELL HAS A BASE FUZE WITH MARKINGS SIMILAR TO THAT SHOWN IN PLATE XV OF THIS PAMPHLET.

(Gr.19. rot Be mit Bd. Z)

Malby & Sons, Lith.

PLATE XIV

—FUZE MARKINGS (TYPICAL.)—

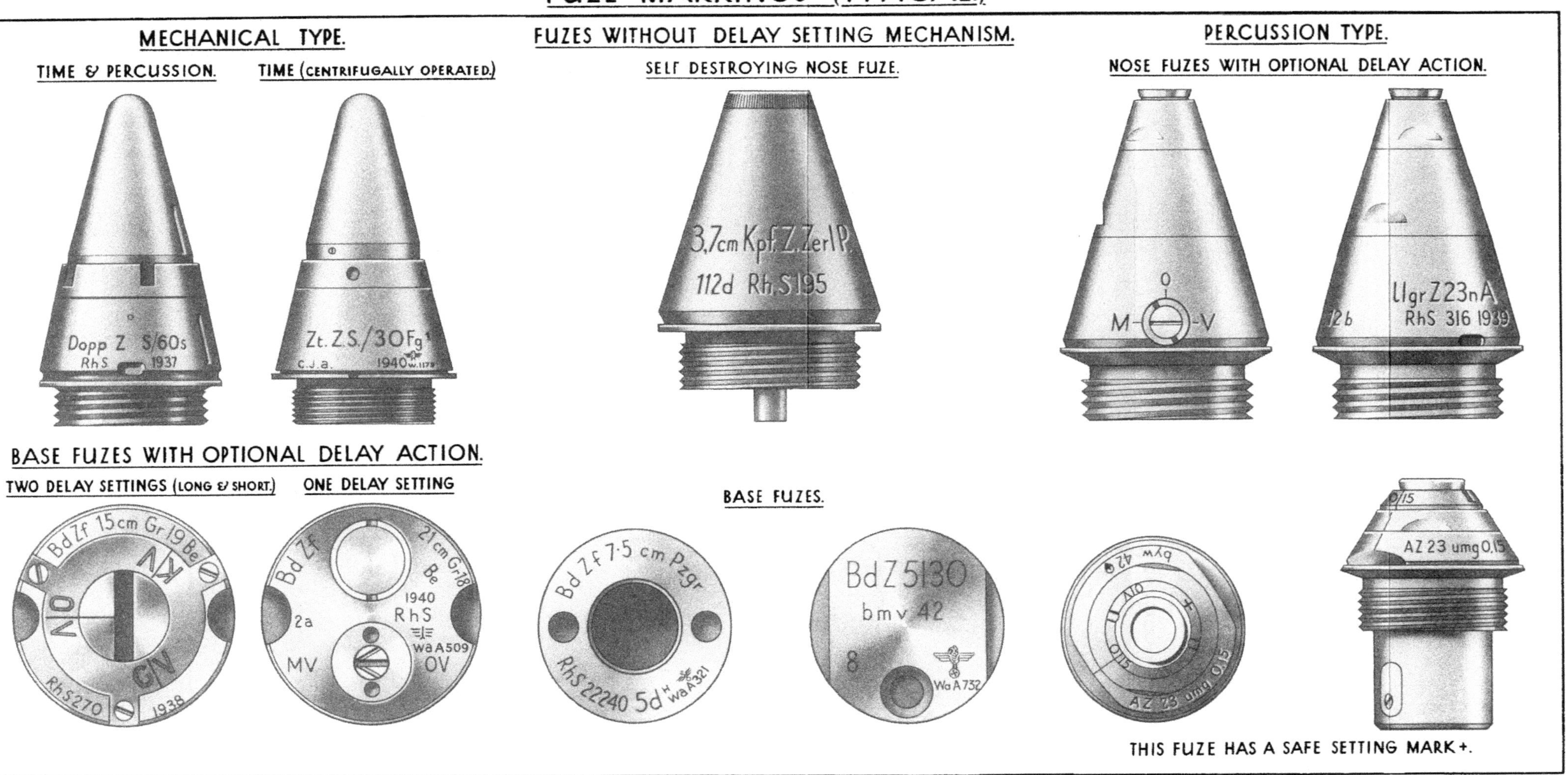

Malby & Sons, Lith.

SMOKE BOX (Typical Markings).

Rauchent W.Nr.9
Phos.
Bitterfeld.
5.L.39.
Cwg. (A)7 Lf.39

Rauchentwickler No.8.
Phosphor
Lief.154.
Sht 1941. Rate 121

GAINES (Typical Markings on side of Gaine).

Gr. Zdlg. C/98 Np.

Kl.Zdlg. 34mV.

Typical Labels.

Zdlg 40
B

Red Band.

Zdlg 41

L.2525. Malby & Sons, Lith.

Plate XVII

PACKAGE LABELS.

Q.F. CARTRIDGES.

FIXED.

1-8,8cm Pzgr. Patr. 39-1 KwK 43 / Pak 43 / Stu K 43 | FES
Bd. Z. 5127 | | 92
Gu. R.P. - G 1,5 - KN.(725/650. 5,1/2)
ktz 1943/10 | GU
Kzu 10.7.43 Da | 6388 St

SEPARATE.

10.5 cm. Geb. H 40
3 Hulsenkartuschen
Digl. Bl. P.-10,5 (10-10-02) Lfg. versch
Gu. Bl. P. A.O.-4-4-06
Lfg. Versch.
Unters. U. Zus. ges: lt.
GU
6327 St

SHELL.

F.H Gr. (Buntr) FES
Dopp ZS/60 Fl *
Gewichtklasse I
Jt. 22.1.44 K.

ROCKET.

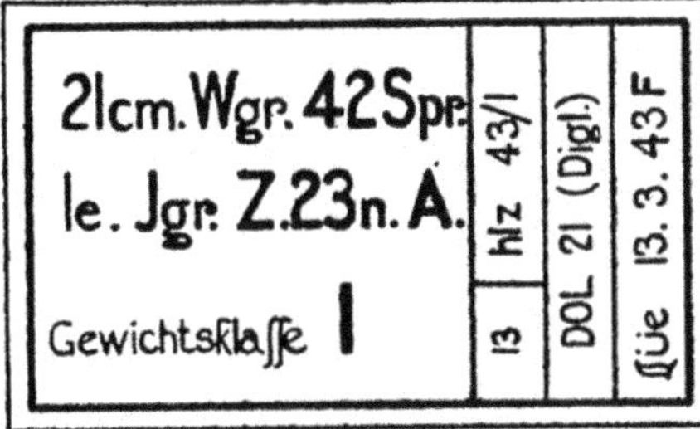

ANTI-TANK ROCKET.

8·8cm R Pz B. Gr 4312.
AZ 5095/1 | Digl. Pre Bl
Digl Pre Bl. 10,5-(100-22/10-14-1,0)
rdf. 1943/9
Tpn. 29.2.44 D.
DR

MORTAR.

8cm. Wgr. 39 umg. | III
Wgr. Z. 38 C | 5 Ldg. | 13
Aw. 16.8.43 D

GRENADE

Sprenggranatpatrone K.P.
Lieferfirma - dag. Lieferung
Angelfertigt - 6.43 40

N.B. The above are copies of some typical labels, to illustrate the information given on them.

Plate XVIII

CODE LETTERS ON PROPELLANT CHARGES AND LABELS.

CODE		MEANING.
D	–	DIGLYLCOL CHARGE.
DR	–	DIGLYLCOL TUBULAR CHARGE.
DST	–	DIGLYLCOL STRIP CHARGE.
GU	–	GUDOL CHARGE
NG	–	NITROGLYCERINE (CORDITE.) CHARGE.
NZ	–	NITROCELLULOSE CHARGE.
DV	–	IMPROVED DIGLYLCOL CHARGE.

www.ingramcontent.com/pod-product-compliance
Ingram Content Group UK Ltd.
Pitfield, Milton Keynes, MK11 3LW, UK
UKHW052106270726
14058UKWH00005BA/660